A New Scientific View of Nature
EXISTENCE

Objects interact and evolve, but first of all, they *exist*. Without Existence, there would be neither Motion, nor Evolution, and there would be no Physical Reality.

Parmenides: "All is Being" (480 BC)

Dr. Alexander Yabrov
Author of the best book of the Year in Health Area
Awarded by the American Publishers Association

Printed by CreateSpace, 2012

Printed by CreateSpace, 2012

ISBN: 1-4774-6217-1
ISBN-13: 9781477462171

Dedication

This book is a tribute to John Archibald Wheeler—dean of American physicists, who formulated the problem of Existence in his famous question: "How come existence?"

Epigraph

"We are missing something Big"
Lee Smolin

CONTENTS

ACKNOWLEDGEMENTS

I use this opportunity to emphasize my profound gratitude to the following scientists.

B. P. Konstantinov—an outstanding physicist—Vice-President of the USSR Academy of Sciences, Director of the Leningrad Physico-Technical Institute. With Konstantinov's support, I was allowed (in spite on the severe non-scientific restrictions) to participate in the contest for a position of a Senior Scientist of the Physico-Technical Institute (then the Institute of Nuclear Physics) of the USSR Academy of Sciences.

In the West, I have greatly benefited from the support of Sir Karl Popper—a world leading philosopher of science and a physicist.

For years, I have collaborated with John Archibald Wheeler—Professor of physics of the Princeton University and Kenneth William Ford—Director of American Institute of Physics. Their constructive criticism and persistent encouragement presented tremendous creative help for which I am profoundly grateful.

I express my deep gratitude to Professors Shoichi Yoshikawa, Department of Astrophysics of Princeton University, Igor Frenkel—Mathematics Department of Yale University, and Mihael Slonim, Department of Electical and Computer Engineering, BenGurion Unuversity, Israel, and Eugene Polsik of the Bohr Institute of Physics, Copenhagen, Denmark. These scientists reviewed my results both in biology and physics. Their constructive criticism, advice, and encouragement played important creative role in the formulation of ideas.

Preface

This book is the first scientific study of Existence.

Currently, the prevailing scientific view of Nature is mechanistic-evolutionary. Natural phenomena are being considered primarily as the manifestations of physical interaction of objects, and also evolution. The new view is different: First of all, objects *exist*. As a matter of fact, we observe different phenomena—Existence, Motion and Evolution—guided by different fundamental processes—those of adequate functioning, interaction, and evolution. The book describes Existence as an independent fundamental category—a leading *state* of the objects. It answers the questions posed by Aristotle: "What is Being?", and by Wheeler: "How come existence?"

Leaning on the shoulders of giants

Look at the circle of thinkers who studied Existence throughout millennia.

Figure 1. Idea of Existence throughout the millennial history of Western thought

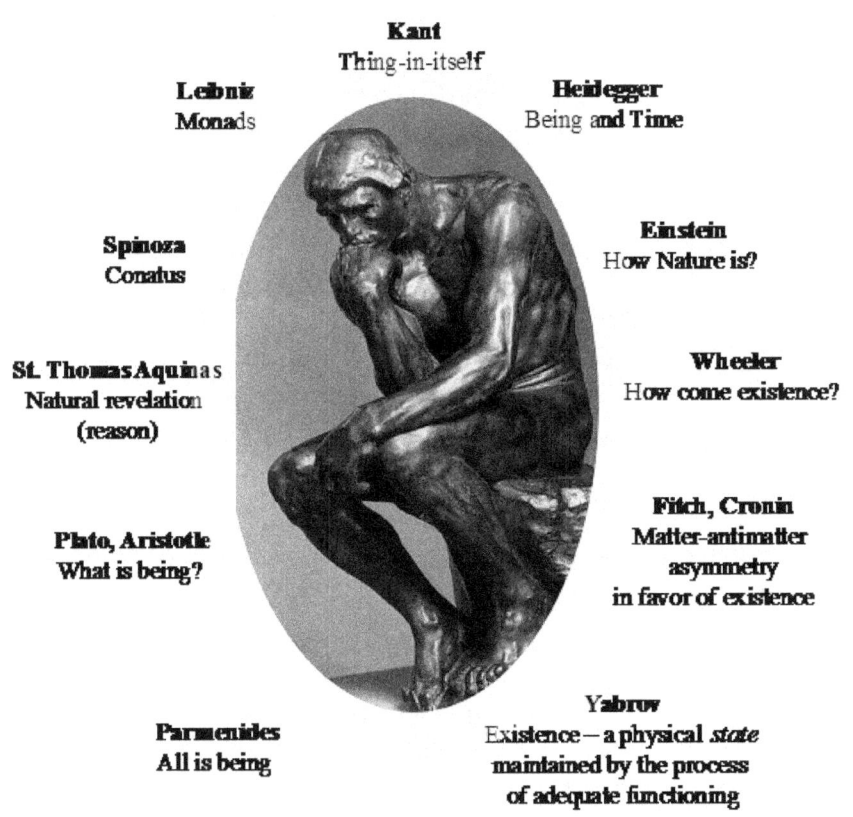

Kant
Thing-in-itself

Leibniz
Monads

Heidegger
Being and Time

Spinoza
Conatus

Einstein
How Nature is?

St. Thomas Aquinas
Natural revelation
(reason)

Wheeler
How come existence?

Plato, Aristotle
What is being?

Fitch, Cronin
Matter-antimatter
asymmetry
in favor of existence

Parmenides
All is being

Yabrov
Existence – a physical *state*
maintained by the process
of adequate functioning

The study goes from the ancient to modern philosophers; then Einstein introduces a *scientific* generalization:

"All these endeavors are based on the belief that existence should have completely harmonious structure. Today we have less ground than ever before for allowing to ourselves to be forced away from this wonderful belief" (1934).

This idea is reflected in the above "circle"—"How Nature is?"

Same year, 1934, Wheeler formulated the scientific problem of Existence in his renowned question: "How come existence?" He studied the problem all his creative life—but did not solve it. In my view, though Wheeler was close to the solution, he could not find the answer because he relied upon the theory and methods of Mechanics (Wheeler, 1991; 1996), rather than those of Existence. Fitch and Cronin gave structural explanation (Figure 1 and further). Our explanation is functional.

I have started working on the problem of Existence in Russia in 1955 after my graduation from the Medical School. Our approach: studying existence of the living organisms using the theory and methods of Biology and Medicine. The studies resulted in discovery of the *process* of Existence of organisms—that of *adequate functioning* (1979). Then I has found that the *same process*——was responsible for existence of the *inanimate* objects. Thus a *new fundamental process*—the one of Existence of natural objects—was discovered (1979; 1986; 2001). Prior to our discovery, two fundamental processes were known: Interaction (Newton) and Evolution (Lamarck, Darwin). Our results were generalized in a theory of Existence, which described new laws of Nature—the Laws of Existence (further and also Yabrov, 1986; 2001).

At those years, because of their central common interest, both groups of researchers united their efforts. Wheeler thoroughly analyzed my results. He enthusiastically approved the direction chosen by me. He was interested especially in the possibility of unification of the inanimate and animate Worlds. Discovery of the *process* of Existence, its *mechanisms* and the *laws*—valid for both kinds of phenomena—allowed me answering the question asked be Wheeler: "How come Existence?" As well as the Aristotle's: "What is being?" (see Table 1).

Introduction

In 1924, Wheeler asked: "How come existence?" The question did not attract attention it deserved: most of the physicists where busy solving puzzles of the quantum mechanics. But Wheeler considered finding the answer to his question being the central problem of science. He looked for the answer during his entire creative life, yet he did not find the solution. In his search, Wheeler was guided by the Bohr's concept of reality. He used methods of the quantum mechanics. Existence, however, cannot be explained based upon the theory and methods of Motion. The *entire view of Nature* should be *different*. It must be a view of Existence rather than Interaction. Guided by this new view of the World, we have solved the problem of Existence. The reader will discover the *theory* and the *laws* of Existence. Among other things, the latter demonstrate that the laws of physics are not the sole Laws of Nature.

Creative collaboration

As described further, I came to the problem of Existence as a result of my studies of living organisms, in particular my medical studies in humans. This work was performed in the USSR. I was not familiar with Wheeler's work, and did not know his famous *question*. I just have realized (after 25 years of my scientific research) that I was studying *existence*. This is what a physician and a biologist-experimenter does: he studies *existence of organisms*.

Thorough analysis of the results led me to a generalization. Discovered is a new area of natural phenomena—those of Existence. It is an independent fundamental characteristic, or state of Nature. It is different from Motion and Evolution—two areas known so far. My further research allowed uncovering that the *animate and inanimate* objects both *exist* by the same fundamental process. Thus the fundamental process of Existence has been discovered. I named it the process of *adequate functioning* (Yabrov, 1979; 2001). In the USA, I have closely collaborated on the problem of Existence with Professor Wheeler.

Is knowledge of physics sufficient for the study of Existence?

A question might arise, whether a physician—researcher not having physicist's educational background—is qualified answering Wheeler's question? Physics proved to be the leading discipline studying Nature as a whole. I explain the commanding position of physics among other disciplines by that it studies Motion, which constitutes our current dominant view of Nature: we see natural phenomena, first of all, as the manifestations of physical interaction of objects. But I have *changed* our fundamental view of the World. As it is described in the book, all natural phenomena are being considered, first of all, as the manifestations of Existence. This refers both to inanimate and animate objects: things and organisms exist. To interact (as well as to evolve), the objects should exist. We study Motion of objects which exist; we study evolution of objects which exist, existed, or will exist. Without Existence there would be neither Motion, nor Evolution, and there would not be Physical Reality.

Nature is not divided by the professional disciplines. Different scientific specializations were introduced by the investigators in the course of development of science—to simplify the work of researchers. Each specialty considers certain area of natural phenomena using different vocabularies, philosophies, theories and methods. The point is, however, that Existence is a subject that cannot be limited by knowledge of a sole scientific discipline. It studies and explains all natural phenomena proceeding from a different (from those of Motion and Evolution) view of Nature—that of Existence of objects.

As described further, even such giants of physics and philosophy as Einstein and Wheeler were—could not solve the problem of Existence, because they based their research upon Motion. As we have discovered, however, Existence *subsumes* Motion and Evolution as its manifestations and mechanisms (Yabrov, 2001).

What kind of knowledge is necessary for the study of Existence?

Knowledge and understanding of physics are necessary but insufficient for the study of Existence. Thorough command of the Life

Sciences—biology, medicine and the social sciences is equally needed. But most importantly—the researcher should not be consumed by a worldview dictated by one's specialty. Today, biologists view life phenomena primarily as the manifestations of Evolution; while physicists think first of all of Motion. Being a physician, I was not subdued by the forceful guidance of the currently dominant views of the World. Doctor is guided by the theory and practice of Medicine rather than those of mechanics or biology. Medical knowledge and experience are the most appropriate for the discovery and study of Existence: a physician studies Existence. It took me quarter of a century of diligent clinical and experimental work to realize this truth (Yabrov, 1969; 1979; 2001). Yet, knowledge of medicine is not enough. Being a Senior Scientist of the Institute of Experimental Medicine, USSR Academy of Medical Sciences, I became a biologist-experimenter specializing in virology and cytology. Then, as a Senior Scientist of the Nuclear Physics Institute of the USSR Academy of Sciences, I acquired knowledge and experimental experience in physics, in particular the quantum mechanics. Add to these my social experience. I grew up without farther. My father spent 18 years in the camps of the GULAG. Both of us—my father and I—did not have doubts that the same fate expects me: he advised that I should drop out of school and acquire profession of a welder, which is needed in the camps. But I went on to become a Doctor—a profession also needed in the camps. I have used the first opportunity to emigrate. Experience of living and working consecutively in four countries having different languages, customs, political systems, and the ways of scientific research—broadened my philosophical horizon, and enriched social and professional experience. As we see, a considerably broader professional knowledge and social experience proved being necessary for a Doctor to discover a new view of Nature, and answer Aristotle's and Wheeler's questions.

PART 1.
VIEW OF NATURE
—A CONCEPT

Decisive role of the views of Nature in development of human knowledge

We have introduced a notion of a fundamental view of Nature *as a currently dominant concept about natural phenomena and Nature.* Discovery of a new fundamental view of Nature is a key event in development of human knowledge. Only three major *scientific* views of nature were discovered to this day—mechanistic, evolutionary, and the quantum mechanical. The scientific views of Nature not only describe—they explain Nature by discovering the underlying Processes, their Mechanisms, and the Laws. Each of the new views of Nature resulted in tremendous advancement of science and technology, and promoted prosperity. Their creative influence persists till today.

Consider this concrete example. Today our dominant view of Nature is that of Motion. Accordingly, we see all natural phenomena primarily as the manifestations of Motion, and the interrelations among natural objects as a result of their physical-chemical interaction (see Yabrov, 2001). This refers both to terrestrial and celestial objects. In other words, we see the World in *motion*. The view of Motion guides our studies and explanations of phenomena. Thus it promotes a certain aspect of our knowledge. Today we consider it to be the apex of scientific achievements. Some researchers suggest that we are close to a complete understanding of natural phenomena and Nature—at least, in principle. Some unresolved problems remain; still dreams about a final theory are being discussed (Weinberg, 1992).

Dr. Alexander Yabrov

Our idea of *different* fundamental views of Nature which follow in succession, each determining our current understanding—does not support the opinion that the above well-established description of Nature is close to final, most complete picture of the World. Moreover, it *contradicts* this opinion of the majority. It becomes clear that the description and understanding of the World is determined by the dominant view of Nature, which changes. A new fundamental view of Nature is discovered—the one of *Existence* of Objects and Nature. It is described in this book. There is no doubt—it will further advance human wellbeing (see also Yabrov, 1979; 1986; 2001).

Dynamics of development and succession of the fundamental views of Nature

Our studies of the subject led us to the following discoveries. Every view of Nature has its stage of an "increasing return", which gradually transforms into the one of progressively "diminishing return". At the latter stage, the current view brings very little new knowledge in spite on tremendous creative efforts. Then it is being succeeded by a new view of Nature and the cycle repeats itself. Another essential trait is that the new views *do not negate* knowledge accumulated by the previous ones. Each new view is a product of the previous knowledge—this is how the treasury of the overall human knowledge is growing. The above findings allow us to consider dynamics of development of general knowledge of mankind as a growing spiral consisting of coils of the views of Nature. This schematic description reflects the transition of the stages of each view from those of increasing to diminishing return; and the succession of the views growing up from one another.

The initial fundamental view of Nature—religion

The subject is so important that I find it useful for the reader to analyze the dynamics of development of the known fundamental views of Nature according to their succession. The initial view of Nature was *religious*. It embraced the period of accumulation of human knowledge starting from origin of the species Homo sapiens during ancient times (lasted 100 thousand years, or more). It had its stage of growing return.

Dynamics of religious views of Nature

Polytheistic religion considered natural phenomena as being unrelated. It idolized some noticeable objects, or events—a mighty river, or an animal, or lightning, or thunder and ascribed them superior powers. This view tore the world in to pieces. People did not even think of a united World.

Monotheism initiated by the Judeo-Christian religion discovered and revealed for the minds of the believers a *one indivisible World* presented via the idea of a *one God*. This was the world of Order. Man was placed in the center of it. This position imposed responsibilities dictated by Morals. Ten Commandments represented apogee of the stage of a *growing return* of the religious view.

Presently, religion is actively opposed by most of the public. Religious views happened to be in the stage of *diminished* return. Religion is considered as being a depository of old superstitious views not supported by the scientific data. Furthermore, it is viewed as a destructive ideology, which propagates global terrorism and anti-Semitism. It is profound mistake, however, to pile in one hip different meanings of the notion of religion, as well as the different religions.

Religion as a source of initial knowledge

We discuss here role of religion as a *source of initial human knowledge* about the World. All the disagreements on whether there is (or there is no) God are irrelevant to our analysis of dynamics of the views of Nature. Judeo-Christian monotheistic religion did not just replace the many-Gods of mythology of polytheism with a one God. This religion *discovered a united, orderly World*—which nobody saw. And it discovered the code of common morals. These discoveries are valid today. They represent the base of our scientific views of Nature, and of our views of adequate human behavior. With time—when our concept of the views of Nature described in this book is understood and accepted—the current attacks against religion: removal of the 10 Commandments from public places, prohibition of the inauguration of students, of treatment of addicts, and the like—will be equated with the book burnings (see also Yabrov, 2001; 2002).

Dr. Alexander Yabrov

The Natural-philosophical view of the world

We characterize the religious view as the one of *Being*, primarily—being of Man. Religious view gave origin to the *natural-philosophical* view (6th—5th centuries BC). The latter also was a view of Being, but it was broader then religious: It included also consideration of Nature. Parmenides (480 BC) declared: "All is Being". Period of *growing* return of the philosophical view was related primarily to the teachings of the ancient Greek philosophers—Parmenides, Plato, Aristotle and others. Philosophy has developed a method of abstract thinking, logical analysis of natural phenomena, and generalization of observations. This method proved being necessary and irreplaceable for the progress of knowledge. Yet, gradually philosophical view slipped into the stage of *diminishing* return, in which it is abiding currently. In its midst, a new kind of views of Nature was developing—the *scientific* ones.

Dynamics of the *scientific* views of Nature

The scientific method of accumulation of knowledge, in addition to observation, was characterized by an active intrusion into natural phenomena by way of experiment helped by mathematics.

Motion

The first area of scientific studies happened to be Motion. It was initiated by Copernicus' discovery of Motion as a fundamental notion *independent* from Being (1543). I consider this time as the beginning of era of modern science (see further and also Yabrov, 2001). The *mechanistic* view of Nature entered its stage of *increasing* return marked by the names of Galileo, Kepler and Descartes. An apogee of the mechanistic view was the discovery by Newton of the theory and the laws of mechanics—described in his "Philosophiae Naturalis Principia Mathematica" (1687). It is necessary to emphasize that philosophy represented an unalienable component of basic science. Descartes developed a mechanistic philosophy, which, by his opinion, explained all natural phenomena. Guiding role of philosophy has been preserved through most of the history of science. Philosophy was named the Quinn of Sciences. The mechanistic view reigned for hundreds of years. It resulted in gigantic advancement of mechanical industry. The latter, in its turn, allowed a decisive melioration of human condition via mechanization of labor.

Evolution

Yet, mechanics could not explain the life phenomena. At the beginning of the 19th century, Lamarck discovered the process of evolution. "Time is insignificant and never a difficulty for Nature", said Lamarck. "It is always at her disposal and represents an unlimited power with which it accomplishes her greatest and smallest tasks (*Hydrogeology*, 1802).

The *evolutionary* view of Nature has started from study of diversity of the living organisms. It reached its apogee of *increasing* return—half a century later, when Darwin discovered the true mechanisms of evolution—random mutations, heredity and natural selection. Darwin developed evolutionary theory, which he described in his "Origin of Species by Means of Natural Selection" (1859). The evolutionary view allowed us to understand the origins of diversity of living creatures. It promoted advancement of agriculture and certain areas of medicine. Till today *all* life phenomena are explained by the evolutionary theory. As we have discovered, however, Evolution of species is *not the only* fundamental principle of Biology. Another fundamental principle is *Existence of organisms* (Yabrov, 1979; 2001; 2012). Current forcible attempts of explaining the every day *existence* of organisms by the evolutionary theory—contradicts the facts of Life. Analysis of innumerous studies in the area of Life Sciences led me to conclusion that currently the evolutionary view is at a far advanced stage of *diminishing* return (see further, and Yabrov, 2001; 2012).

Quantum mechanical view

The mechanistic view guided our understanding of Nature for more than 200 years. It remains valid today. Yet, at the end of the 19th century, it entered its stage of *diminishing* return. It could not explain newly discovered phenomena—electron, x-rays, radiation, and others. These discoveries happened to be the beginning of a new scientific view of Nature—the *quantum mechanical* one. Its stage of *increasing* return lasted three quarters of the 20th century. It provided for tremendous advancement of basic and applied physics. It is enough to mention the achievements in the areas of information and communication. Yet, at the end of past century, the quantum mechanical view has entered the stage of *diminishing* return. "A growing number of theoretical physicists see the present situation as a crisis of fundamental physics discovering the laws of nature"—says Smolin in his book "The Trouble with Physics" (2006). By his opinion, the creative recession is a consequence of change of "methodology and style of research": "The great physicists of the beginning of the 20th century—Einstein, Bohr, Mach, Boltzmann, Poincare, Schrödinger, and Heisenberg—thought of theoretical physics as *philosophical* endeavor." This opinion is in a complete agreement with our analysis of development of knowledge via the successive views of Nature. Philosophy should necessarily be applied in the fundamental scientific studies. Scientists-philosophers worked at the stage of increasing return. But lately, "shut up and calculate" became a mantra, emphasizes Smolin. This is when the quantum mechanical view crossed the threshold of *increasing* return toward the *diminishing* one. Time has come for the change of the leading view of Nature. Says Einstein: "Changes of view are continually forced upon us by our attempts to understand reality" (Einstein, Iinfeld, 1937).

Summary. Our analysis of the dynamics of development of knowledge allows the following conclusions. Human knowledge develops via succession of the fundamental views of Nature. Of these we start from the religious and philosophical—abstract views of Being. They were succeeded by the *scientific* ones: Mechanistic, Evolutionary and the Quantum Mechanical. All fundamental knowledge of mankind is the product of these views of Nature. Currently, our knowledge is fragmented because interrelation-interconnection between the views is distorted. The last scientific view tends to dominate our entire understanding of Nature—pushing aside the achievements of the previous views and hindering development of the following one. This is illustrated by disregard of the role of religion, skeptical attitude to philosophy, and resentment to any unconventional basic ideas. This tendency impoverishes the treasury of human knowledge. We come to conclusion that all the above views are in a stage of diminishing return. The time has ripened for a *new* fundamental view of Nature.

The new view is presented below. As did the previous fundamental views, it broadens further our understanding of Nature. Moreover, it advances and interconnects all previous fundamental views. As a book describing a new fundamental scientific view of Nature, this volume could be compared to the "Principia" by Newton and the "Origin of Species" by Darwin, which is a powerful stimulus for the reader.

PART 2.
ON THE ROAD TO
A NEW VIEW OF
THE WORLD

The ascent on Everest

I am a mountain climber (Caucasus—Georgian Republic, and the Krasnoyarsk rocky pillars—Siberia). Comparison with mountain climbing is very appropriate here. When you start the ascent—you do not see the summit—it is hidden in the clouds. In case of the scientific studies—you often do not even realize the precise aim of your investigation. This was said by Einstein: "If we knew what it was we were doing, it would not be called research, would it?" (1996).

You do the experiments. You discover new particular phenomena. You compare them with the known data. Gradually, a direction of further studies looms—it leads upward. The higher, the more difficult is the advancement. But at the same time—the broader your vision becomes. You realize the interrelations and interconnection of facts,

which previously seemed scattered. New knowledge promotes your advancement. Eventually, you overcome the wall of the upper clouds and reach the summit. This is the highest stage of a scientific investigation. Your ideas soar. (This expression—K. Ford, Director of American Institute of Physics, who is a pilot—applied inscribing for me his book "In Love with Flying" (2007): "To Alex Yabrov, whose ideas sore"; this is how I became acquainted with this word).

This is the stage of discoveries of the concepts, theories and the laws. Then, your generalizations could be used by other explorers in their studies. This is how science advances.

The above description fits the dynamics of the scientific studies of the problem of Existence. It is comparable with the ascent on Everest. Three teams—two American—led by Einstein and Wheeler, and one Russian—led by Yabrov, have started in different times, using different routes to the top.

Einstein

This was Berlios who said: "Bach is Bach, as God is God". Yet, it took centuries for Bach's works being fully understood and appreciated. We perceive an analogous situation with the works by Einstein. It is not broadly publicized in the popular scientific literature, but it is well known to physicists that all the studies by Einstein starting from his dispute with Bohr and afterwards are considered being a complete failure. Einstein new it—the discrepancy of the views was irreconcilable. In his *Reply to Criticism* on the occasion of his 70es birthday he wrote:

"I discovered that the mentality which underlies a few of the assays differs so radically from my own, that I am incapable of saying anything useful about them" (see Schlipp, p. 665, 1959)

In spite of the rejectionist attitude of his colleagues, Einstein continued his persistent creative search.

This Chapter is not an attempt to justify Einstein's work. He does not need it—he is *Einstein* in spite on the opinion that after 1925 "he could go fishing" (see Hawking, Penrose, pp. 134-135; 1996).

My purpose is to try to explain what actually occupied Einstein's thoughts during all those years of hard creative (unacknowledged) labor. Understanding should help us to reveal the direction of his search for harmony in Nature.

I return to the analysis of Einstein's disagreement with the dominant generalization of the probabilistic view of Nature. I explain Einstein's difference with the views of the majority as follows. The majority of physicists consider Nature from the worldview of Motion. Whereas Einstein's view was *broader*—he saw natural phenomena as the manifestations of Existence of Nature. His goal was to find out "Ho nature is?" However, being a physicist, Einstein considered only the phenomena of Motion. Here we perceive a profound paradox. As it follows from our studies, Motion is a *mechanism* of Existence (further, and also Yabrov, 2001). Phenomena of Motion *are* the *manifestations* of Existence of objects at the levels of physical particles and bodies. Similar to Newton and to the ancients, Einstein thought of *Existence* of Nature, I suggest. Seeing Existence in the phenomena of Motion, Einstein felt the narrowness of the views of most of his colleagues and therefore could not agree with them. Popper describes that in their personal discussions he called Einstein—*Parmenides*, and Einstein agreed with this characterization (Popper, p.90; 1983). Parmenides, as we know, considered all phenomena of Nature as the manifestations of Being (Existence).

I suggested that Einstein's *dilemma* of explaining natural phenomena in general is comparable with that of Wittgenstein. Wittgenstein saw something crucially important, but he could not convey *what* it was. On our view, Wittgenstein *saw* phenomena of Existence and discerned them, but he could not express his vision being limited by his mechanistic worldview (for more details, see Yabrov, 2002).

Einstein "felt" the necessity of a different, broader view of Nature (see further). And yet, and yet—Einstein did not come up with the concept and theory of Existence. His remaining on the position of a worldview of Motion kept Einstein at the level of his opponents. Furthermore, his position was disadvantageous, since he *saw more* than he could prove and explain—using theories and methods, which were not sufficiently appropriate for his embrace of the breadth of the problem under study—Existence of Nature.

Einstein was right in his broader view of Nature. In particular, he was right in his famous dispute with Bohr—denying the principle of uncertainty a *universal* status. The essence of the dispute of titans remains to be the central problem of physics. I found it so important that allotted a separate book to this problem. Our concept of *relativity of uncertainty* allowed us to prove Einstein being right in his dispute with Bohr (see Yabrov, 2012a).

Great discoveries of the previous years have opened before Einstein's eyes the broadest panorama of *existing* Nature. Vision of this all-embracing perspective prompted his search for its scientific explanation—his ascent on Everest. His tantalizing search for explanation of harmony in Nature positioned Einstein ahead of his time. Yet, the limiting influence of a conventional view of Motion prevented him from reaching the summit: to discover Existence.

The Contribution by Wheeler

Modern philosophy has negated the problem of Existence as a "nonsensical". B. Russell suggests a simplistic interpretation according to which "exists" is equated to "is"—"This clears up two millennia of muddle-headedness about 'existence', beginning with

Plato's *Theaetetus*" (B. Russell, 1961). This simplistic interpretation of a profound philosophical-scientific notion is a reflection of the stage of diminishing return by modern philosophy, I believe (Yabrov, 2001).

However, the problem of explaining Nature as a whole was revived by *science*. A unique role in its rekindling has plaid John Archibald Wheeler of Princeton—a discoverer of such notions and terms as black holes, geon, quantum foam, wormholes, and others.

The essence of Wheeler's crucial contribution into philosophy and practice of physics is reflected in his questions. And he actively participated together with the others in finding the answers. Among these questions are: "How come Existence?", "It from Bit?", "Why the Quantum?", "A participatory Universe?", "What makes Meaning?", and others.

Of these questions, the major and all-embracing one is "How come existence?" (Wheeler, 1996). By asking this question, Wheeler restored and re-approved the might of the penetrating vision of thoughts by Parmenides, Plato and Aristotle. He was one of the few who understood Einstein.

Wheeler was the leader of second American team searching for a new view of Nature.

This is how Wheeler (1991) describes his "own still-unended straggle with the age-old issue" of how come existence:

"Three legs mark this sixty-year zig-zag odyssey. Each sailed for a different star: 1. All is electrons (1934- 1950); 2. All is spacetime continuum (1953-1973); 3. All is observer-participancy (1973—)". (p. 5, 1991)

Wheeler concludes the description of his studies of the problem of Existence with an optimistic prediction expressed in the words of Bohr: "Tomorrow is going to be *wonderful* because tonight I do not understand *anything*". Thus Wheeler acknowledges that he did not find the answer to his question. My explanation why he did not reach the summit is analogous to that suggested by us to the work by Einstein. Thorough analysis of Wheeler's own description of the stages of his search—quoted just above (1991), show that Wheeler remained within the realm of Motion.

Wheeler—Copernicus—discoveries of an equal scale

As described below, Copernicus discovered Motion as a fundamental notion independent from Being. His discovery paved the way for those by Galileo, Kepler, and Descartes. And it served as a conceptual base for the theory of mechanics by Newton. Wheeler—first in the world—touched upon the problem of Existence. By posing his question "How come existence?"—Wheeler has pointed out importance of a scientific understanding of how it takes place (Wheeler, 1991; Wheeler, 1996; Wheeler, Ford, 1998). He realized that knowledge of motion could not answer his question.

Favoritisms of fate

Products of basic science—concepts, theories, laws—are the objective results of observations and experiments. Science, however, is a one of manifestations of the human social activity. Inevitably, the subjective factors—personal views, preferences, preconceptions, and the like—play role in the fate of a scientific discovery. The more original and all-embracing this discovery is—the more difficult is its acceptance.

Hard fate of Wheeler's discovery

Wheeler, first in the World, broached the subject of Existence. By my evaluation, this discovery equaled by its embrace the achievement by Copernicus. Wheeler's discovery deserved being qualified as a scientific revolution. Yet nobody has noticed it. The only scientist who realized the true importance of the discovery—was Wheeler himself. He did not abandon his belief. It took a half of century while another scientist—*working independently on the opposite side of the globe*—developed the idea of Existence as a fundamental category independent from Motion and Evolution, and explained how it takes place (Yabrov, 1979; 2001).

Fortunate fate of the de Broglie's discovery

Wheeler posed his question at 23. The year was 1934. At that time everybody where occupied with the problems of quantum mechanics. The Einstein-Bohr dispute was at its pick. It seemed that no other basic topic could attract attention of physicists. Unexpectedly, young Wheeler introduced his question of How come Existence. It did not attract attention of the colleagues.

Another scientist of precisely the same age—de Broglie—10 years earlier, in 1924, introduced an idea that the *material* particles possessed a wave. Prominent Professors—among them the curator of de Broglie's dissertation—dismissed the idea as having no sense. The only scientist who said that the idea actually did have sense—was Einstein. His support happened to be sufficient to influence the opinion of the colleagues. The concept of wave-corpuscular nature of material particles was accepted. Furthermore, it was extrapolated upon the physical bodies. In three years, de Broglie was awarded the Nobel Prize. The wave theory became an inalienable part of the quantum mechanics.

History of these two fundamental discoveries is a dramatic illustration of the role that fortune plays sometime in development of science

PART 3.
OUR STUDIES OF EXISTENCE

I approached the ascent from a different route—*independently* from the American investigators. I have started my studies as a practicing physician, immediately after graduating from the 1st Leningrad Medical Institute—a medical school of a European stature. My specialties: internal and infectious diseases. Medicine is a particularly creative field. Every person is unique. Any disease manifests itself differently in a certain patient. Doctor is experimenting continuously. For example, I came to conclusion that a child with tonsillitis or scarlet fever, not having traces of pus in the throat—should not be treated with antibiotics. If, however, I see even only the dots of pus—I apply an antibiotic. In both cases my concern—to avoid complications of the kidneys, which might appear a month later. Now, 50 years after I have published my results (Yabrov, Gutkin, 1958), antibiotics are used too often without individualization. Then, I became a biologist-experimenter. Areas of studies: viruses and cells. Primary subject—mechanisms of resistance; these mechanisms are responsible for survival of both the healthy and the ill. The principal method of study of the life micro-objects, such as viruses and cells, is experimental. This situation is similar to that in the quantum mechanics. Gradually, based upon analysis of the uncountable experiments and also clinical investigations, I came to understanding that I was studying *Existence of living organisms*—not Evolution and not Motion (1969; 1979; 1980).

I have presented a compendium of my results to Sir Karl Popper—world renowned philosopher and theoretician in both biology and physics. I spoke of Existence as fundamental category of natural phenomena, and its underplaying process. As a result, a close creative interaction developed between us, which lasted many years. Popper enthusiastically supported my view of Existence. He was a member of a scientific Board of the international journal "Medical Hypotheses". He introduced me to the editor recommending publishing my materials. The article describing the *concept of Existence* and the *fundamental process* underlying phenomena of Existence—that of *adequate functioning*—was published in three months (1979). Then followed the next paper, which described the *mechanisms r*esponsible for the process (1980). I have also published a book "Interferon and Nonspecific Resistance" (1980a), which included material on Existence.

In the following five years of intensive experimental studies, to my astonishment, I came to conclusion that the *inanimate* objects existed by the *same* process as the *animate* did. Since I have discovered a fundamental regularity characteristic both for animate and inanimate objects, I claimed it to be a Law of Nature. I presented an article to the Medical Hypotheses. The article was published (1986).

Thus I have advanced by the slope of the Everest. I have emigrated from Russia in 1974 when it was a very risky endeavor: you never knew—where you would end up—in Siberia or abroad. I have followed my studies in different countries—Russia, Israel, Canada, and the USA. Working in different conditions, using new methods, new approaches and models essentially promoted my advancement. During the 1980s, I have resolved most of the questions related to the explanation of Existence of both animate and inanimate objects. I developed a concept and theory of Existence; discovered the process, its mechanisms and the Laws (see further). The results were published in a series of papers (1979; 1986; 1987; 1987a; 2001a; 2004) and monographs (1980; 2001; 2002; 2012).

We come to conclusion that the Russian team *conquered* the summit of the Everest. Our success was achieved because—in difference from such outstanding researchers as are Enstein and Wheeler—we were guided by a *new view of Nature*—that of Existence, discovered by us.

The Wheeler's formulation

At the time of my work on Existence, there were no studies on this problem anywhere in the world—that would be known to me. In the US, I reside in Princeton. From the materials of a scientific conference at the Princeton University, I found that Professor Wheeler was working on this subject. Then I have learnt of Wheeler's formulation of the problem of Existence from his books (Wheeler, 1996; Wheeler, Ford, 1998). I was startled. It was clear to me that this scientist was a genius.

Reading the question for the first time, I realized instantly that Wheeler—first in the world—suggested that Existence should be studied. It was clear *to him* that this was impossible to explain how everything existed based upon the current knowledge of that time. He always emphasized that the answer represented the *central problem of science* (Wheeler, 1996). His broad investigations in various areas of physics, for which he became famous, were—in fact—reflections of his persistent search for the answer to his question.

Schopenhauer said: "Talent hits a target no one else can hit; genius hits a target no one else can see". The situation with Existence is different. Everybody *sees* that things exist. But very few think of Existence as a *discrete* independent fundamental notion needing a special study. Here fit rather the analogous utterances by Schrödinger and Szent-Gyorgyi—Nobel laureates in physics and biology: "The task is not to see what has never been seen before, but to think what has never been thought before about what you see everyday."

This is the point—it was necessary to *think* of *Existence*. Wheeler did it first in the world. And so did Yabrov in his independent studies in Russia.

I have undertaken all my efforts to meet with Wheeler—to show him my work. We have met at the beginning of 2000s—after publication of both my books "How Man Exists" (2001) and the "Tractatus Scientifico-Philosophicus" (2002). After the initial familiarization with the published results of mine, regular meetings at the Wheelers office at the Jarvis Hall in the Princeton University were arranged. I was bringing my materials piece by piece. We discussed them with Wheeler. Wheeler took with him some parts—they were discussed at the next meeting. Wheeler mostly approved the materials. Every his commentary was taken into account. These evaluations resulted in a very thorough and comprehensive analysis. This was a real creative collaboration. Some of my friends–scientists expressed amazement that this giant dealt with me at all. It was incomprehensible that the physicist of the highest world class perceived an interest in the writings of a physician-biologist. But this was the central point—Wheeler was especially interested in the problem of unification of the inanimate and animate Worlds based upon the concept of Existence.

I find it appropriate to describe some details—to convey to the readers the atmosphere of enthusiasm of our meetings. Wheeler always greeted my appearance exclaiming: "Doctor Y-a-a-brov!" (I was always amazed that he remembered my unusual name). And he was leaving his chair—to help me with my coat (gentlemen's school);   I hectically tore off my jacket.

This was a collaborative work of the highest creative strain, indeed. It sublimated decades of experimental studies and their scientific-philosophical generalizations. This unification of the individual achievements helped me to come to a scientific generalization of the broadest possible scale—formulation of a *new* all-embracing scientific view of Nature described in this book.

Spiral of scientific knowledge and of the human knowledge in general has grown by its next coil. As did the previous general views of Nature, the one of Existence shall advance science and promote wellbeing of mankind, I believe.

My retrospective analysis of our collaboration led me to the following thoughts. We *both* were longing for this meeting and for collaboration. I have had a distinguished support from Sir Karl Popper and David Horrobin (editor of Medical Hypotheses) when I have discovered the process of existence of organisms and showed that it was also valid for inanimate objects (1979; 1986). But this was quarter of a century ago. Since then I met nothing but misunderstanding and resentment. For me, collaboration with Wheeler was the first encouragement and acknowledgement after decades of toiling in isolation. This is the fate of every scientist who dares to introduce unconventional ideas. Though it might seem paradoxical, Wheeler remained in isolation—as far as the central subject of his thought was concerned—during all his creative life. And now a Russian biologist has proved Wheeler being right all these years. Mutual rapport has appeared instantly—we understood each other from half a word. In a one of his letters, Wheeler wrote:

"Dear Dr. Yabrov, now that we have had a chance to spend some time together, I enjoyed the opportunity to read your materials on the general theme of existence. I find it provocative—especially your focus on seeking a united view of the animate and inanimate worlds. Your writing has the flavor of the finest ancient philosophers...Sincerely, John A. Wheeler". February 17, 2005.

Precisely, philosophy of an increasing return (see above) is necessary in order to come to a new scientific view of Nature. Being-Existence is a central millennial problem of mankind. Wheeler keenly realized it. He was first who gave it a scientific start (his classical question of science: How?). Science and Philosophy unite when the fundamental scientific problems are being studied. This union is provided by the scientists-philosophers—the leaders, to whom Wheeler deservedly belongs.

A personal comment. In this and in our previous books on the problem of Existence, my name appears together with the names of Wheeler, Aristotle, Spinoza, Bohr, Einstein, Darwin, and other outstanding researchers of Nature. It might seem that I compare *myself* with these legendary minds. I have thought of it in the process of my generalizations. For a while, I felt confused, and this hindered the progress of my work. A serious analysis, however, led me to a conclusion that I do not compare *myself* with the other researches. What is being compared—ideas, notions, concepts and theories. This comparison is necessary in order to elucidate whether progress has been achieved.

PART 4.
A NEW VIEW OF NATURE IN A NUTSHELL: CONCEPT, EXPERIMENTS, THEORY AND THE LAWS

The CONCEPT

Conversation with the reader

As a result of tens of years of an indefatigable experimental work—both exiting and monotonous, with its ups and downs, we have reached the summit. Now our task has changed. Our aim is a fundamental generalization. Together with the reader, we should glide through the broadest space which the notion of Existence embraces. In the process of our creative flight, we should discover and formulate the concept, the theory and the laws of Existence.

Dr. Alexander Yabrov

Role of a general scientific concept—example: concept of Motion

    A concept plays central initiating role in any broad creative undertaking. A scientific concept formulates the main idea and defines the scope of phenomena under the study. It unifies the efforts of many investigators who belong to different professions and use different methods. It is an *arch* which bears the meaning of the whole construct. Thus, for example, Francis Bacon (1561-1626) in his *Novum Organum* (1606-1620), developed a concept of modern science as an active, organized, systematic inquiry equipped with a certain method aimed at the study of Nature. Below, a comparative analysis of the concepts of Motion and Existence is given. This comparison, based upon familiar examples, introduces the reader into the Philosophy and Theory of Existence.

Copernicus—Notion of Motion Independent from the Notion of Being

    As an example close to our study, we should consider briefly the development by Copernicus of the concept of Motion.

    Copernicus (1473-1543), a Polish astronomer, introduced a theory of the heavens whereby the Earth revolved around the stationary Sun. Copernicus' views differed from those of those Biblical and Aristotelian by which the Earth was considered the Center of the cosmos. The Copernicus' contribution to the treasure-house of human knowledge is classified as revolutionary. The accepted explanation for this highest evaluation is that the Copernican theory removed the Earth, and hence man, from the center of the universe.

    It is necessary to emphasize that the idea that the Earth rotates and revolves around the Sun was put forward by Aristarchus of Samos (310-230 BC)—the Greek astronomer also known for his calculations of the size and distances of the Moon and the Sun; though important, his ideas remained within the realm of astronomy. The Copernican theory, however, required a complete change in man's philosophical conception of the universe. We propose that the essence of the Copernican revolution should be interpreted *differently* proceeding from our concept of the fundamental notions (see below).

The fundamental revolutionary difference between Aristotelian and Copernican astronomies is that Aristotle described Being, while Copernicus described Motion.

Aristotle's description of the universe as a tangible thing characterized by its physical structure and by the function of perpetual circular motion of the "First Heaven" around the Earth is in full accordance with the notion of Being. Aristotle described the existing universe as he perceived its being, and he saw it's functioning as analogous to his description of the being and functioning of the human body and its parts. In Aristotle's descriptions of celestial and terrestrial phenomena, motion was part of Being. For example, Aristotle characterized development of a plant from a seed as motion.

In difference of the conventional explanations, I assert (Yabrov, 2001) that Copernicus discovered Motion as an *independent* fundamental notion. Figure 1 is a Copernican dynamic scheme of the universe consisting of imaginary lines and circles showing the paths of the planets.

Dr. Alexander Yabrov

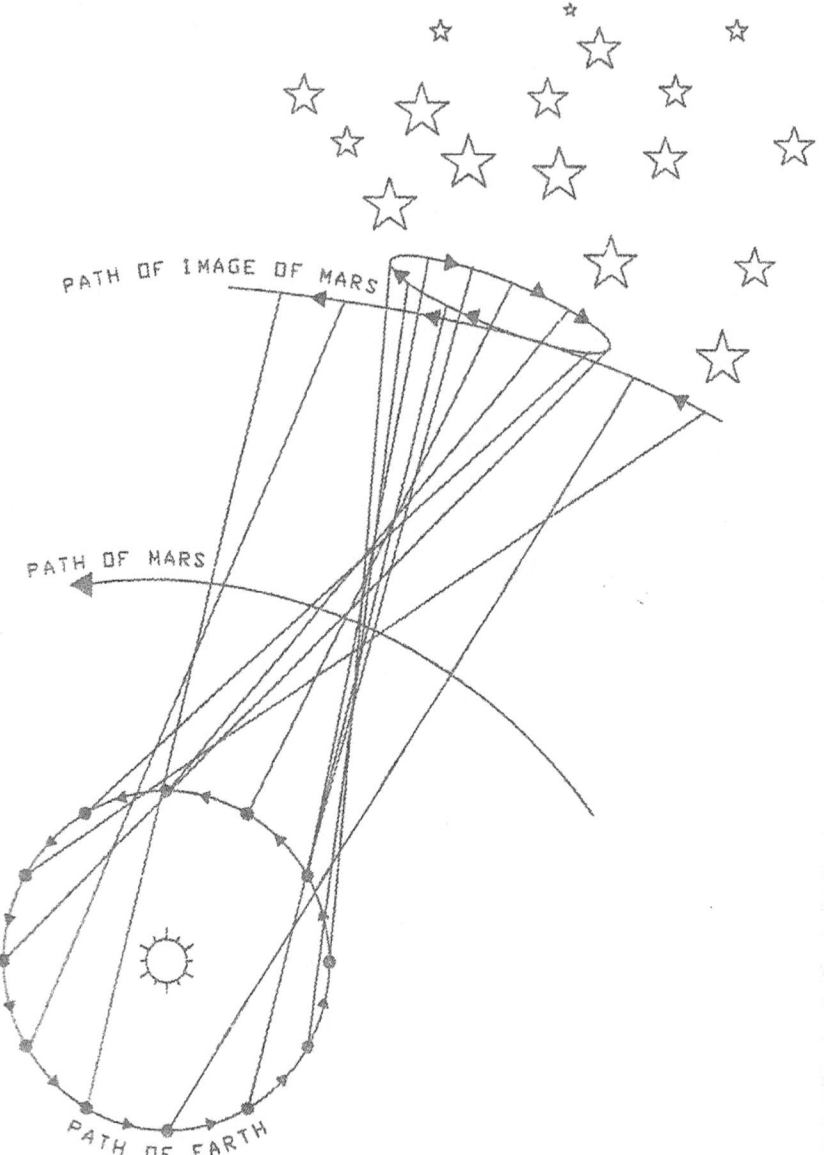

Figure 1. Copernican Explanation of Retrograde Motion (from D. Park, 1988, p.144).

The planets are depicted as mere points repeated sequentially in order to present the *path* of their *motion*. Gone are the "First Heaven"

24

and the planetary spheres. The whole construction, the thing that was perceptible, palpable like the bones and flesh of the body has disappeared!

It is replaced by the description of the *motion* of the celestial bodies.

Copernican physics, mathematics and astronomy are the sciences of Motion, which he uses for description of motion.

Yes, Copernicus substantiated the idea of the Earth moving around the Sun. However, the heliocentric system would remain an important scientific theory in a particular area, viz.—astronomy. But besides this discovery, though important, still limited by a particular field of study, Copernicus also presented to the world a general one: that was his discovery of Motion as an independent (from Being) fundamental notion, or an independent characteristic, or function of Nature. It became clear that the phenomena of nature could be viewed, described, and explained from another view, different from that of Aristotle's.

This is exactly what came to pass. The perspective on things and the phenomena related to them changed, and the view of the world too—shifted. Thenceforth natural phenomena were considered as manifestations of Motion. This is because his discovery changed the fundamental notions that determined the new outlook on Nature, the Copernicus contribution deserves to be recognized as a revolution, I suggest. The science of motion flourished, and its influence spread beyond the realm of science proper. A new view of the World, a *mechanistic* one, was formulated by Descartes, based upon a new fundamental notion of Motion.

The Copernican discovery allowed one to distinguish natural phenomena related to Motion. However, in the era initiated by Copernicus investigators were interested not only in observing phenomena, but in finding an explanation of *how* the phenomena occurred.

Based on a new concept, philosophers and experimenters undertook joint efforts in understanding phenomena of motion. Important contributors were Galileo (1564-1642), who applied mathematics to the study of motion and proposed the law of uniform acceleration for

Dr. Alexander Yabrov

falling bodies; Kepler (1571-1630) who discovered the elliptic shape of the orbits by which planets moved around the Sun, and described his planetary laws; and Descartes (1596-1650), who advanced further the mathematical method, and developed a mechanistic worldview.

One might wonder why the works of these researchers so well-known and highly appraised are not considered revolutionary. The answer becomes clear based on the concept of fundamental notions (Yabrov, 2001). We classify the discoveries of the *fundamental notions* and *theories* as revolutions. This refers to the Copernicus' discovery and then the discovery of the theory of classical mechanics by Newton.

We believe that the example of development of the *concept of Motion* illustrates that a concept is a necessary starting point of a broad scientific study.

Fundamental *notions* describing Nature. On the threshold to the new concept

"How to describe Nature as a whole?"—I posed the question to myself. This was probably how the ancients probed Nature.

The answer came in stages. First, without long ruminations—effortlessly—came an idea of the "fundamental notions".

Observations of natural phenomena of various kinds and analysis of experiments in animate and inanimate objects led me to a generalization that they all are the manifestations of Motion, Origin and Existence, or of their combination. There are *no* phenomena *beyond* theses notions—none!—These notions describe the entire Nature. Therefore we named them the *fundamental notions* (Yabrov, 2001).

Comparison with the ancients

Generalizations of natural phenomena by the Western civilization could be found as far back as those by the ancient Greeks. Parmenides thought that all was Being, immobile and unchanging, whereas Motion and Change, in Parmenides' view, were just an illusion. The atomists, Leucippus and Democritus, said that everything was the result of Dynamics of indivisible atoms in a void. Heraclites spoke of a continuous flux, or Change. There were also other generalizations describing Nature. For example, the notion of the *mean* among the opposites, e.g., between hot and cold, bitter and sweet, white and black; and also the consideration of Soil, Water, Air and Fire as the basis of all phenomena.

Contemplations in this direction reveal a connotation with these ideas of the ancients. This is natural, since the objects of observations remain the same, i.e., the phenomena of Nature. The subject of thought is the same, namely—how to describe and explain all what we observe. But there are *differences*; and they are not just semantic.

The ancient philosophers described all the phenomena by a *one* notion. Thus, Parmenides' Being covered the whole of Nature. Using terminology employed in this study, it included phenomena covered by the notions of Existence, Origin, and Motion. This is how I interpret his principle "all in one"—Parmenides described all manifestations of reality by a single notion. Democritus considered *all* natural phenomena as manifestations of the dynamics of atoms. Whereas Heraclites saw *everything* as a reflection of change.

As follows from our study, such a generalization of all the phenomena under a sole notion prevents an understanding of *how* the phenomena of nature actually occur. Phenomena occur according to certain *processes*. No a one and the same process exists that would underlie *all* the phenomena. If we unite all the phenomena by a single concept of Being (or Dynamics, or Change), we cannot suggest a common process by which they occur and, therefore, remain unable to explain how each group of phenomena takes place. This obstacle disappears when we base our vision of reality upon our fundamental notions. Each one has *its* domain of phenomena because the processes, which underlie them, are different: phenomena of Motion proceed by the process of physical-chemical interaction; those of Origin by the process of evolution; and those of Existence—by the process of existence, or the process of adequate functioning (see further).

The other difference is closely related to the former. The scientific concepts of Motion, Origin, and Existence carry an idea of a necessity of their *explanation* as the pertinent phenomena are understood today by *science*. They imply the objects, the processes, and the mechanisms involved in the observed phenomena, i.e., our fundamental notions direct an investigator toward the search for answers to the What *and* the How. Whereas general notions by the ancient philosophers were viewed as explanations in-themselves: complete and final—answering What.

It should be emphasized that many particular concepts, which were introduced by the ancient philosophers, are being used by modern science, for example matter, force, acceleration, void, and many others. These concepts are applied because they precisely convey the meanings used by the contemporary investigators to describe pertinent natural phenomena.

I consciously do not follow this path, and introduce different terms in order to emphasize that our fundamental scientific notions of Existence, Origin, and Motion have *their* profound meaning, *each* describing *its* largest area of natural phenomena, and forming a solid foundation for a scientific study—via discovery of the underlying fundamental processes and their mechanisms.

Concept of the fundamental *States* of Nature. State of Existence—a leading state

It was said: "Physical concepts are free creations of human mind..." (Einstein, Infeld, 1960).

The next major step in development of our general views about Nature was the *reformation* of the idea of the fundamental notions into the concept of the *Fundamental States of Nature*. This metamorphosis was not just a semantic exercise. It was a *new* concept incited by the necessity to answer the *How*. The task of Philosophy—to *describe* Nature—answering What. The task of Science—to *explain*—answering both What and How. Our discovery that Motion, Origin (Evolution) and Existence are in fact the physical *states*—is in itself a decisive step toward the *scientific* explanation and eventual understanding of physical reality as a whole.

An expression "state" is applied in many situations. We speak of a "serious state of affairs", of a "state of elation", of a "state of art", of a "gaseous state", and of many other states—spiritual and physical, which characterize different objects, under different circumstances, and in different times. Our concept of the "States of Nature" considers the *physical states* in which *all* natural objects, inanimate and animate, and Nature—abide *all the time*. Thus the concept of the "fundamental states" *reifies* an abstract philosophical notion of Nature, or Reality and thus converts it into a materialistic scientific concept of Physical Reality (Dictionary: *reify*—[from L.—thing; and—*fy*—to treat as substantially existing, or as a concrete material object).

State of Existence—a leading characteristic of physical reality

Motion and Evolution are the established scientific concepts explained by the fundamental scientific theories of mechanics (classical and quantum) and evolution, correspondingly. Therefore their consideration as the *states* of Nature is easily acceptable.

Now we introduce an idea that there is *another* general physical state that characterizes condition of all natural objects and Nature— that of *Existence*. It is a necessary characteristic of reality of natural objects: *they exist*. Furthermore, this is a *leading* characteristic. The states of Existence, Motion, and Evolution, embrace and describe all natural phenomena. But the state of Existence is the central one in our description and understanding of reality. Consider this. We study motion (interaction) of objects, which *exist*; we study origin and change of the objects, which *exist, existed* or *will exist*. Without the *state of existence* of objects, there would be *neither* motion, *nor* evolution. And there would be *no* physical reality.

The following considerations illustrate the State of Existence being a central characteristic of natural phenomena subsuming those of Motion and Evolution as its *manifestations and mechanisms*.

Motion is a mechanism and a manifestation of Existence

Further, in our discussion of the *mechanisms* of Existence, an idea is introduced that the existing natural objects of different complexity are composed of the structural-functional levels of organization that are common for all objects (The 2nd Law—Table 1).

The Table will be discussed in more details later. Here I would like to attract attention to the atomic-molecular level and that of the physical bodies. An analysis shows that at these levels motion participates in the maintenance of *existence* of objects together with the forces of interaction (see also Yabrov, 2001). For example, existence of the solar system as a natural object is maintained via gravitation of the planets toward the Sun balanced by a continuous motion of the planets by the certain orbits. It is correct to say that motion and gravitation are the *mechanisms*, which maintain existence of the solar system—this was what Kepler and Newton had discovered: how the solar system *exists*. Motion also participates together with the other forces of interaction in the maintenance of existence of an atom. Thus it is correct to say that motion is a *mechanism* responsible for *existence* of the solar system; the same refers to *existence* of an atom (Yabrov, 2001). Similar consideration is valid for the objects of any complexity including organisms, because their existence is provided by the *concerted* interaction of mechanisms acting at all the levels of their organization. Logically (though somewhat unexpectedly) we come to conclusion that the State of Motion *provides for* the State of Existence at the structural-functional levels of the atoms and the physical bodies. The following consideration is apt here:

"Throughout two hundred years of scientific research force and matter were the underlying concepts in all endeavors to understand nature. It is impossible to imagine one without the other because matter demonstrates its existence as a source of force by its action on other matter" (Einstein, Infeld, p. 52; 1937).

This utterance illustrates the idea that motion (interaction) is a mechanism and a manifestation (demonstration) of existence. We thus speak of a general state of Existence, which subsumes motion as its mechanism and a manifestation (Yabrov, 2001).

Natural selection is not blind—it promotes Existence
Wheeler speaks of "...the blind dice of mutation and natural selection that lead to life and consciousness and observership at some point down the road" (Wheeler, 1996).

We agree with Wheeler when he says that the *random* mutations are blind. But the natural selection is *not*. It selects the objects, which *exist longer*. For example, in the course of development of the solar system there were formations whose properties were not favorable enough to prevent their leaving the system by flying away, or falling on the sun. Selected happened to be those, which could function adequately and thus exist within the solar system. Now consider the following examples related to the living organisms. In the process of their multiplication, bacteria mutate very often. If we add an antibiotic into the growth medium, most microorganisms die out. But those, which happened to have a mutation causing their resistance to this antibiotic—do not die. Furthermore, they multiply more rapidly having no competition. This is how selection of the antibiotic-resistant strains of bacteria occurs. Another example is related to animals. It should be emphasized that not all the random mutations are favorable for the organism in which a mutation occurred. In some instances mutations are unfavorable for existence of the individual. For example, some fetuses are born with a pathologically small brain, or with no brain at all; or with severe malformation of the spinal cord. These individuals are not selected for longer life—they die in the womb or at birth, or soon after birth. In difference from these, the organisms having mutations favorable for existence thrive and have an enhanced opportunity of having more offspring and thus spread their traits advantageous for healthier, longer existence. This is how natural selection works, or "selects"—rather than killing those less fit—it facilitates existence of the organisms whose adequate functioning is relatively optimal. We come to conclusion that natural selection is not "blind"—it *promotes* Existence. In this sense it plays role of a *mechanism* of Existence (Yabrov, 2001; 2012a).

Thus Existence is the leading fundamental state subsuming Motion and Evolution as its mechanisms and manifestations.

The State of Existence is a physical category

We came to characterization of Existence as a *"state"* having in mind an analogy with such medical terms as a "state of health", or a "state of disease"; we characterize a *state* of a patient as a "stable condition", or a "critical condition". These are not some abstract philosophical notions. These terms describe an actual *physical* state of a person. A similar concrete physical meaning we put into the term "a state of existence of an object"—an object *exists*—this is its State. This understanding implies the *necessity* of explanation of *how it occurs* that the object and Nature as a whole—exist: "How come existence?" (Wheeler 1996). The explanation is provided by a scientific theory (see further).

PART 5.
MODELS,
METHODS,
EXPERIMENTS

Conversation with the reader

The reader should necessarily have in mind that the concept, theory and the general laws of Existence described here are firmly based upon analysis of thousands and thousands of experiments both in animate and inanimate objects. When you are on the summit, it is hard to see the base of the mountain. Yet, this is the way of science: Via the countless experiments to the fundamental generalizations. I find it appropriate to repeat. Discovery of a new fundamental view of Nature: Existence of individual natural objects and Nature as a whole—is a result of decades of observations, experiments, and their gradually broadening generalizations. Aim of this Segment is to familiarize the reader with the actual investigations, which provided a base for the concept and the theory of Existence. Of course, it is impossible to describe here all the pertinent data. For the pertinent articles see Yabrov, 1958; 1961; 1966; 1967; 1969; 1971; 1975; 1977; 1979; 1986; 1982; 1987; 2001; Yabrov and Smorodintsev, 1967; Yabrov and Okunev, 2004; and the monographs (Yabrov, 1980; 2001; 2002; 2012; 2012a). Below, the studies in the living organisms are briefly described.

Dr. Alexander Yabrov

Models

Choice of a model had a decisive role for the successful exploration of phenomena of Existence. Being a physician and a biologist-experimenter, I worked with the living organisms—studying *how they exist*. Multifarious models were exploited. First among these were the humans. Man is the most complex organism and a most complex natural object. Complexity begets variety. Each person is unique. In addition to the general differences of sex and age, people differ in their health, needs, aspirations, circumstances, and other important aspects. An absolutely unique feature of this model is that man can speak. So, the investigator may get direct answers to some of questions.

A polar difference from the human organism as a model for scientific studies—represented the cells cultivated in the test-tubes—outside of the body—in tissue cultures. Besides the human cells—those from various other animals were used. The cells were from healthy, or from cancerous tissues. These models allowed studying the immediate needs of the cells under various conditions of cultivation. They also allowed us to monitor development of a disease from its very start.

Besides the single cells grown in cultures, various organisms of increasing complexity were used as experimental models. The simplest of these: viruses. We studied the broadest spectrum of viruses: from the polio, consisting only of a nucleic acid, to ectromelia virus (smallpox of mice), which possess close to two dozens own enzymes. Thus viruses represented different experimental models in themselves.

Of other infection-related organisms I should name malaria parasites; ticks and mosquitoes—transmitters of encephalitis. We also worked with the cold-blooded—sea animals—mollusks and electrical skates. Then followed the warm-blooded: chicken embryos, mice, rats, hamsters, guinea pigs, rabbits, and monkeys.

Use of different models allowed us, on one hand, studying the individual and species-specific peculiarities of existence of different organisms. On the other hand, the common features that we discovered in so many different organisms allowed us broad generalizations concerned existence of the living organisms.

Methods

We differentiate two categories of a scientific method: a *general* method and a *particular*. Two different general scientific methods are known. The first is that of inductive reasoning (*induction*) developed by Francis Bacon (1561-1626). According to the method of induction, generalizations can be made based only on a thorough investigation of gradually accumulated, well-established facts. Hence the ensuing, broader generalizations are developed from previous limited ones. The whole study based upon precise and sufficient particular observations and experiments gradually leads to a broad generalization explaining how the phenomena under study occur.

Another method is that of *deduction*. Here a scientist *first* makes a broad generalization by suggesting a hypothesis explaining the phenomenon. He then undertakes a study to check whether the hypothesis can be refuted. This is how Feynman describes the deductive method:

"In general we look for a new law by the following process. First we guess it. Then we compute the consequences of the guess to see what would be implied if this law that we guessed is right. Then we compare the result of the computation to nature, with experiment or experience, compare it directly with observation, to see if it works" (Feynman, p. 156, 1990).

At present the deductive method overwhelmingly prevails; induction is considered outmoded.

Of the above two methods, I chose *induction*. It occurred naturally. The sphere I was venturing to explore was a new one. No coherent data to make broad guesses and generalizations where available. As a matter of fact, no scientific data on the subject existed—at all. Contemplating my situation, I realized how Bacon arrived at the method of induction. Bacon developed his method at the dawn of science. He understood that attempts at rapid broad generalizations of scattered superficial observations did not lead to a comprehension of how phenomena occurred. Analysis of his failures led him to the only right approach—that of a thorough accumulation of particular consecutive observations, and their detailed study, followed by *limited* generalizations of pertinent facts. This was followed by the *more broad* generalizations based on a discovery of interrelations, many of them covert, between the limited ones. *Gradually*, a comprehension of a broad phenomenon in its entirety ensues. Darwin describes the method he exploited at the start of his work as follows::

> "My first note-book was opened in July 1837. I worked on true Baconian principles, and without any theory collected facts on a wholesale scale, more especially with respect to domesticated productions, by printed inquiries, by conversation with skillful breeders and gardeners, and by extensive reading." (Darwin, *Autobiography*, 1876)

This is how I worked as a physician and then as a biologist-experimenter—by the method of induction.

In the process of investigations, multifarious *particular* methods and techniques were used. Of those, I find it necessary to name the following.

Various clinical lab-tests. Tissue culture—monolayer and suspension. Microscopy—light and electron. Various methods of cytology, virology, histology and histochemistry, biochemistry, biophysics, physical chemistry, chemistry, physics, and others (see Yabrov, 1980).

Experiments

Profession of a biologist-experimenter—virologist and cell biologist consists in doing experiments—many in a day—all year long. The time for analysis of the results and for reading of scientific literature: the long late evening hours. Contemplations—24 hours. Amount of experiments performed in the course of our study is hardly countable. Moreover, this kind of formal calculation would not make sense. Important are the results—they should bring new knowledge and show the way for further research. We may just conclude that our entire study is based upon unending thorough experimentation. This is not all—we necessarily should add clinical work of a physician. Thoughtful doctor experiments all time during one's career. Thousands of patients—each one having *his* course of a disease depending on age, sex, circumstances, resistance, combination of maladies, properties of damaging factors, heredity, etc. New equipment and new medications are being developed. New diseases appear. To help the patient more effectively—doctor should experiment. Now you have an idea of the role of an experiment in the study of existence of organisms. Below are given some examples.

Immediately after finishing the experimental part of my PhD dissertation (interaction of the non-specific and specific mechanisms of immunity), I have received an assignment to organize and lead a tissue culture Laboratory at the Department of Virology, Leningrad Institute of Experimental Medicine, USSR Academy of Medical Sciences. At that time—late 1950es—the method of tissue cultures was not developed completely; especially in Russia. Standard media for growth of the tissue were not manufactured. There were no disposable plastic dishes, test-tubes and pipettes; only reusable glassware. Everything should have been developed anew—on the premises: Method of mass production of nontoxic distilled water; experimental selection of the optimal ingredients for growth media; as well as of the conditions of cultivation of the cells from different organisms, say chicken, monkey or human. Even dishwashing demanded special study: The cells did not survive if the thoroughly washed glass dishes remained contaminated with the traces of washing chemicals, while the letter was almost impossible to get rid of. Conditions of the work where barely sustainable: We worked literally in the boxes (sterile): sealed rooms, no air conditioning, strong UV lamps on—all the time (protective broad-rim hats, and gaggles); add to these gauze masks, latex gloves, and permanent hit from gas burners. All these measures were necessary against contamination of the cultures with microbes. This exhaustive, seemingly strictly manufacturing experimental work brought creative results, eventually. Using cultures, I have isolated a strain of respiratory sincytial virus from a child with pneumonia—first in the USSR: A serious lung infection of children and elderly (Yabrov, Golubev, Smorodintsev, 1964). This is far from being all, however. Persistent experimentation with the cultures of cells isolated from different animals and sleepless nights of contemplations about optimal conditions of their survival allowed me to look at the cells anew. This was the essence of my work to find out how the cells exist beyond an organism: Find their needs, select favorable conditions, and the like. Gradually I started to realize that the human cells were *organisms* in themselves, having their needs for survival and the mechanisms to maintain their existence. It might seem trivial but at that time and to these days, doctors, pathologists, and even cytologists view a cell merely as a part of the body. Consider this. Most of the clinical tests study how the cells fulfill the needs of our body: sugar and cholesterol blood levels, blood pressure, and others. But, as we found—function of human cell is dual. To survive—it should satisfy both its *own* needs *and* the needs of the *body*. Imbalance of satisfaction of these needs is the starting point of a chronic disease. As briefly described in this book, we have developed a new theory of pathology (N) and suggested new methods of prophylaxis, diagnosis and treatment of chronic diseases (see further—Consequences). Theory of pathology applied to this day has been introduced 150 years ago. It does not consider function of the cells—only their structure. To develop a new general theory of pathology is a serious contribution. It was achieved as a result of our experimental studies of the tissue cultures (Yabrov, 1980; 1987).

The leading subject of our investigations was: The mechanisms of resistance. These are the mechanisms, which determine the outcome of a disease, and of existence itself. The more we know about these mechanisms, the more effectively we may help the patient. Among such mechanisms is interferon—an antiviral protein produced in our body in response to viral infection (Isaacs, Lindenmann, 1956). In the early 60es, the American and French publications appeared showing that interferon was produced also in response to bacterial toxins. It is my principle that in biology, a phenomenon, which repeats itself, does not occur by chance: you should search for a *mechanism* (Yabrov, 1980a). I have presumed that interferon protected organism not only against viruses, but also against *bacterial* products. In a special series of experiments, we have developed a technique for measuring damaging effect of bacterial products (toxins, endotoxins) in the tissue cultures (Yabrov at al., 1967). Using this experimental model, we have shown that interferon *did* increase resistance of cells to bacterial products. These experiments, first of all, demonstrated that interferon protective effect of interferon was not limited by viruses only. At the same time, our results provoked a question about the mechanism of protective action of interferon. The fact that interferon protected cells against a *non-multiplying* agent was discovered for the first time. Prior to our experiments, interferon was considered inhibiting the multiplication, of viruses and also of the cancer cells. But bacterial toxins did not multiply.

Based on analysis of various experimental data, I came up with a new hypothesis of protective action of interferon. We have suggested that interferon was not a specifically antiviral or anticancer factor, but rather a mechanism of *maintenance of fidelity of the cell protein synthesis:* It prevented synthesis of abnormal proteins. As such, interferon protects the cells against different agents, which could damage protein synthesis—whether viruses, carcinogens, toxins, or others. To test this hypothesis experimentally, I have selected non-multiplying damaging agents known to impair cell protein synthesis. Among them: a physical factor—radiation, and a chemical one -antibiotic streptomycin (known to damage protein synthesis in both bacterial and the human cells). The letter experiments were especially difficult and sophisticated. We traced dynamics of impairment of a certain protein (myeloperoxydase) in the cells under the influence of streptomycin—in presence (or absence) of interferon protection. Both—in the case of radiation and the antibiotic—we proved the fact of protection (Yabrov, 1980; 1982; Yabrov, Sverdlov, 1969).

Thorough analysis of our results let me to the idea that interferon represented a *special kind of the mechanisms of resistance.* Usually, we view defense mechanisms as acting against certain damaging factors. Consider example of antibodies. Their protective effect is directed against certain damaging agents (antigens)—bacterial, viral, and others. In case of interferon, however, we are dealing with the mechanism whose protective effect is directed *not* against a particular damaging factor, but rather toward the *maintenance* of a certain vital structure-function of a cell. In case of interferon—maintenance of fidelity of protein synthesis—independently on the character of the damaging agent. Are there other mechanisms of resistance of this kind? Comparative analysis brought me to conclusion that interferon was not the only mechanism of resistance of this kind. Another example is the DNA repair system. It maintains another vital structure-function of the cell—the genome. Still another example of a new kind of mechanisms of resistance represents the mechanism protecting the *energetic structure-function*—discovered by us (Yabrov, 1975)). Based upon these experimental data, I have made a generalization that our body possesses a *special group of the mechanisms of resistance* whose action is directed toward maintenance of the vital structures-functions of the cell. I have named them—*mechanisms of security* (Yabrov, 1980; 1980a). I have presented a Proposal—to study these mechanisms to the Health Department of the White House. To my amazement, I got a personal call from Director of the newly created Office of Home Security—Mr. Tom Ridge (it happened 10th of October, 2001—a month after the 9-11 terrorist acts). Mr. Ridge informed me that the Office was planning to invite me to Washington to discuss the details of the project. He emphasized that I was right—this should have been a long-term research project. Unfortunately, the Office soon happened being overwhelmed with acute antiterrorist problems and my Proposal did not get further development. But the fact itself demonstrates importance of our studies of the mechanisms of resistance (this episode is described in details in Yabrov, 2002).

Still another example of implementation of our experimental studies. I have found that the special preparations of hydrolysates of the animal proteins—prepared by us—essentially increased resistance of the cells to different viruses. Both the DNA and RNA containing viruses were sensitive to the protective effect. Resistance to diphtheria toxin (a highly potent damaging agent) was also increased (Yabrov, 1971). It is a very rare observation of having broad nonspecific protective effect against various viruses and diphtheria. We have undertaken broad experimental investigation. For reason of space, I would prefer to limit presentation by description of some original results. In the process of studying the biochemical mechanisms of protection, we have discovered that certain components of our preparations possessed powerful regulatory influence upon cell respiration.

A one drop of our material completely revitalized the cells whose mitochondria had stopped functioning. Further analysis allowed us to include these substances of natural origin in the group of the mechanisms of security maintaining cell energy structure-function (see above and Yabrov, 1975; 1980a).

It was interesting to find out whether an effect found by us takes place in the organism. After extensive research, I have chosen the sea animals as our model for these experiments. Some of these organisms—sharks, electrical skates, and mollusks are known to possess an increased concentration of protein catabolites in their blood. Our experiments have confirmed that these naturally produced substances demonstrated capacity of increasing resistance (Yabrov, 1971; 1976). These results led me to the following generalization. It is a usual view that in the organism—regulatory substances are produced as a result of the process of synthesis—e.g., enzymes, interferon and others. Consider, however, the following well known facts. Intensive exercise and fasting result in activation of body functioning. I hypothesize that the invigorating regulatory effect is caused in these situations by the products of *catabolism* rather than synthesis. Add to these examples healing effect of the so called "wound hormones". Discovery and study of the regulatory role of the products of catabolism opened to us a new kingdom of natural modulators of metabolism—*potential drugs* (see Yabrov, 1980a). For example, recently, it was published that a traumatic brain injury can spur a massive development of new neurons in the brain of mice (Requarth, Crist, 2011).

Add to these also the results of experiments by other researchers—used in our study. We come to conclusion that our study of Existence of objects is based upon voluminous experimental data.

PART 6.
DEVELOPMENT OF AN APPROACH TO THE THEORY OF EXISTENCE

The aim of this Part—to prepare the reader to the Theory of Existence.

Difficulties of presentation of a new view

Today, the mechanistic, evolutionary and quantum mechanical views of Nature are broadly acknowledged and accepted. But at the time when each of these views was introduced—they were confronted with skepticism and in many cases were dismissed by the majority. Very few realized the importance a new view from the start. Bohr said that Einstein and science were lucky that his initial article read Planck; "any other editor would through it away into a paper basket". The worldview of Existence is not an exception.

Our debate with Prigogine

This is an illustration of a creative discrepancy of the views concerning the completeness of our current knowledge of Nature, my opponent being a prominent scientist—Nobel Laureate—whose contribution in the areas of physics and chemistry are broadly acknowledged.

Dr. Alexander Yabrov

In summer 1998, by my request, I had a two-day personal discussion of the problem of Existence with Ilya Prigogine . I came to Brussels at his Solvey Institute (Brussels)—specially for this meeting. We both agreed that the problem was important, but from that point our views diverged.

Yabrov's view: All the manifestations of Nature are described by the notions of Motion, Origin and Existence, the latter being the leading one. Aim of the meeting—to hear criticism of the scientific theory explaining *how* objects exist.

Prigogine's view: Existence as such is a "superfluous notion—everything is described by the process of evolution of the moving particles—this is how everything exists".

Vigorous debate did not draw the participants to any compromise. A home reading of some materials from the forthcoming book (Yabrov, 2001)—what Prigogine kindly suggested—did not change the opinions. Inability to convey the essence of the idea was a terrible frustration (this meeting is described in detail in Yabrov, 2002).

On my return to Princeton, I received a FAX:

"Brussels, 18th June, 1998. It was a pleasure to meet you in Brussels. I can only say that I wish you the best success in your endeavour. Sincerely yours, I. Prigogine".

The example of our debate with Prigogine illustrates that even the leading scientists consider the current evolutionary-mechanistic view of Nature to be maximally comprehensive, rejecting any broader view as being superfluous. It also shows that it is very hard to formulate a new fundamental concept and substantiate it by a corresponding theory being confronted by an established conventional view. This is why I have comparatively few publications. The broader a new idea is—the stronger is resentment.

The following example is pertinent.

Initial opinion of Professor K. Ford

A new scientific theory needs evaluation. In many cases the judges are seasoned, profoundly knowledgeable researchers having their established views and preferences. Subconsciously, such a specialist reads the material deviating from one's views—with a dismissive attitude, or at least with an aggravated skepticism. Below, I quote Professor Ford—Director of American Institute of Physics, whose opinion I value very highly. Our creative contacts started around the time of the above meeting with Prigogin. When the books on the problem of Existence have been published—Yabrov, 2001; 2002—I asked Ford to have a look, and to show these works to Wheeler. Below is his Internet answer.

From: "Ken Ford"
To: "Alexander Yabrov" Cc: <->
November, 09, 2002
Subject: science and philosophy
Dear Dr. Yabrov,

On Thursday (November 7) I gave John Wheeler his copy of your "How Man Exists" and discussed it briefly with him. He said he would try to read parts of it. I recommended that he concentrate on Chapter 3. After he has examined the book, he may want to suggest that you get together.

Thanks also for the copies of your "Tractatus Scientifico-Philosophicus."

I stand in awe of your intellectual endeavors, yet feel that I reside in a separate universe from the one you inhabit. To me, "How Man Exists" is a work of philosophy, not science. It impressively extends philosophical inquiries into existence that have occupied great minds from Aristotle or before to the present, but it lacks the features that I associate with science—predicted outcomes of experiments and a quantitative description of nature. Perhaps my characterization of science is too narrow—for instance, would Darwin's work fit it? Nevertheless, I just can't agree with your contention that you have moved the study of existence from philosophy to science. Among other problems I have with your work is that I don't find it helpful to seek common principles (or "laws") of existence for pebbles and people. I guess you will have to put me down as a Philistine too mired in conventional approaches to science to be able to appreciate revolutionary advances from a different quarter.

I will have a look at "Tractatus Scientifico-Philosophicus" and send you my reactions later.

Thanks again for sharing these works with us.

Ken Ford.

Dr. Alexander Yabrov

The letter reflects Ford's critical style. It is very important and helpful, but tough. Again, it illustrates the difficulty of presenting a new scientific theory. Below are given some examples of application of a new view for resolution of every day phenomena.

New view explains the everyday phenomena

We start with examples from every day life in order to illustrate that a new worldview and a theory explaining it are necessary, because the current view does not allow us understanding of simple every day phenomena. What makes our examples especially persuasive is the fact that the explanations cannot be found even by such outstanding researchers as Wheeler, Ford and Feynman are.

A question might arise: What for a new general theory is needed? We have a theory of quantum mechanics describing all, or most of phenomena. It is largely acknowledged by the scientists and the lay public alike that we are living in a *probabilistic* world. Based on the quantum mechanical worldview, *all* the natural phenomena are considered from the probabilistic point of view. The following is a standard example. Pieces of a torn in-half paper remain separate. But according to the dominant worldview, these pieces might come together and restore the original sheet, though the statistical probability of this to happen is very low.

"You won't live long enough to see it happen. The universe won't even last long enough for it to happen, yet in principle two torn sheets could weld themselves into a single smooth sheet" (Wheeler, Ford, p. 347, 1998).

The view based on the theory of Existence is decisively different. The pieces of paper *will not* "weld themselves"—this should not occur *in principle*. The statistical probabilistic principle is a reflection of the worldview of *Motion*, whereas most of the every day phenomena are the manifestations of *Existence*. They are underlain and governed by the process of *adequate functioning* and not of physical-chemical interactions. Therefore they can be understood and explained only based on a different worldview—that of Existence (see further, and also Yabrov, 2001).

46

Analogously, we cannot accept the explanation of irreversibility of time by the complexity of the objects.

"The intriguing thought that follows from these considerations is that we are aware of a one-way flow of time only because we ourselves are complex systems interacting with other complex systems. We remember the past and not the future not because there is any fundamental asymmetry in time but because of the overwhelming disparity between likely and unlikely in everything that we are and do see." Then follow the examples from discussions of Wheeler with Feynman "about electrons moving with equal ease backward and forward..." (Wheeler, Ford, p. 348; 1998).

The point is not that the electrons are simple objects. Wheeler and Feynman think of Motion, whereas they consider examples from the area of phenomena of Existence—of "ourselves and other complex systems". We do remember the past because it has *occurred*, and we do not remember future because it *did not* occur. Occurrence (or not) of a phenomenon and remembering it are the notions from the area of *Existence*. It is not a matter of some "fundamental asymmetry in time". This is a matter of the fundamental differences of the areas of natural phenomena under consideration, and of the processes and mechanisms involved (see further).

Feynman, in his lecture *The Distinction of Past and Future* (1990), presents the students a seemingly inexplicable fact that the every day phenomena are evidently irreversible. His example:

"If one drops a cup and it breaks, the pieces do not come together and "jump back" into your hand". The next example he cites to illustrate the irreversibility of phenomena: "When one reverses a section of a movie picture containing a number of phenomena, everybody laughs because the events running in this sequence do not occur in the real world. But if you show any phenomenon involving only gravitation running in on a film the sequence will look quite "normal" and will not evoke laughter" (p. 107).

Feynman lists various physical laws—gravitation, electricity and electrical magnetism, and the laws of nuclear interaction—and

Dr. Alexander Yabrov

concludes that all are time-reversible. This represents a puzzle. Let's solve it.

Feynman's analysis is as follows.

"If the world of nature is made of atoms, and we too are made of atoms and obey physical laws, the most obvious interpretation of this evident distinction between past and future, and this irreversibility of all phenomena, would be that some laws, some of the motion laws of the atoms, are going one way—that the atom laws are not such that they can go either way...But we have not found this, yet. That is, in all the laws of physics that we have found so far, there not seem to be any distinction between the past and the future (Feynman, p. 108, 1990).

Feynman does not give an explanation. He asserts that "laws of physics do not have direct relevance to experience" and he affirms that knowledge of physical laws does not assure an understanding of phenomena.

"In fact, although in these lectures we have been talking about the fundaments of the physical laws, I must say immediately that one does not, by knowing all the fundamental laws as we know them to-day, immediately obtain an understanding of anything much. It takes a while, and even then it is only partial" (p.122, 1990).

The point of the matter is that Feynman suggests that the physical are "all the fundamental laws", knowledge of which is necessary and *sufficient* for understanding of everything. As follows from the concept of the fundamental states, *physical* laws explain only the phenomena of Motion. In order the phenomena belonging to the fundamental notions Existence and Origin be understood, we need to discover and apply *different* laws of Nature.

Analysis of the contradictions

Let us try to analyze the above ideas in the light of the concept of Existence. When Feynman states that the movie picture with "a number of phenomena" being reversed—provokes laughter, he speaks of the phenomena of our daily life. These are phenomena, which belong to the realm of *Existence*. The same refers to the example, of the broken cup. The cup's falling and its fragmenting are two phenomena belonging to different spheres—those of Motion, and Existence, re-

48

spectively. Consideration of Feynman's analysis helps to perceive the difference:

"...the law of gravitation is of such a kind that the direction does not make any difference, if you show any phenomenon involving only gravitation running backward on a film it will look perfectly satisfactory" (Feynman, p. 109, 1990).

The point is that the phenomenon of physical interaction is *mutual*. When an apple falls to the Earth, the Earth also *falls* upon the apple. This is not by chance Feynman states that for the film to look "normal," a phenomenon involving *only* gravitation needs be shown in reverse. In other words, only the physical interaction should be demonstrated—that is only the phenomenon belonging to Motion, so as not to provoke laughter. But a cup that ceased to exist as an object is a phenomenon of Existence therefore it provokes laughter when shown in reverse. We may state that, in *contradistinction* from the physical laws, those of Existence are time-irreversible.

Thus the laws of physics do not explain *all* the complexity of natural phenomena.

We often hear from physicists, mathematicians and computer scientists that, everything can be explained by the physical laws. Nobody, however, has achieved this, yet. It might seem as if physicists do not have sufficient time to consider every aspect of things, because they are always preoccupied with their more important work. Feynman says of the unexplained phenomena: "We have not thought them all through yet".

It is hard to agree, however, that physics did not have sufficient time to solve the problems of existence. Modern science can claim close to a half of millennium of successful studies (starting from discovery of Motion by Copernicus—see above). During this time, physics, which proved its efficacy in discovering the laws of Nature, would have explained the phenomena discussed above by Feynman and Wheeler—*if it could*.

The number of steps lying in between such seemingly unrelated phenomena as the fall of an apple, motion of the Moon and the Earth, and the tidal waves of the oceans—appear to be innumerable. Yet physics found the interconnection and defined it by a corresponding law because the process (interaction) and the mechanism (gravitation) underlying all these phenomena belong to the sphere of Motion, a realm studied by physics.

Instead of waiting for things *somehow* to find their explanation based on the currently acknowledged laws, we should admit that these laws *cannot* provide an explanation of *all* natural phenomena. Not because the physical laws are wrong, but because they are only applicable as an explanation of phenomena relating to Motion. Although it is broad, this area does not embrace *all* natural phenomena. Therefore *new laws* are needed, and the first step to their discovery, is the acknowledgment that phenomena, which are not amenable to explanation by physical laws, belong to an entirely different sphere: that elusive one, which everyone observes, but fails recognizing—being influenced by the dominant mechanistic worldview. Utterance by Schrödinger fits here: "The task is not to see what has never been seen before, but to think what has never been thought before about what you see everyday." Both Wheeler and Yabrov followed this thesis.

Some clarifications

As the reader might notice, we do not rush with the immediate description of the theory as such, which comprises the new key notions, the process, the mechanisms, and the laws. Rather we prefer to depict some pertinent details of the work. In doing so, we pursue a multifarious goal. First of all, we want to involve the reader in the atmosphere of novelty of certain tasks. They appear only when the problem intended for study is entirely original and basic. Here belong the selection of a general method, discovery of the general key notions, and the like. Another task is to clarify some aspects, which might provoke misunderstanding.

The following clarifications should help to remove the seeming incongruities. In this book the adjectives "fundamental" and "basic" are not used interchangeably. These are *different* scientific terms, which characterize notions, theories and processes differing by their scope. We name Existence, Motion and Origin—the *fundamental notions*—to emphasize that they embrace the *broadest possible* spheres of natural phenomena. There are no phenomena beyond these notions-areas. Accordingly, the scientific theories explaining all the phenomena pertinent to each of these areas of Nature we name the *fundamental theories*. Here belong the theories of mechanics, evolution, and existence. A theory can explain *only* those phenomena, which are underlain by a general process discovered by *this* theory (Yabrov, 2001). Therefore, the processes explaining and governing all the phenomena belonging to a certain fundamental area we characterize as the *fundamental processes*. Phenomena of Motion and Origin are underlain and governed by the fundamental processes of physical-chemical interactions and of evolution, correspondingly. The fundamental process governing *all* the phenomena of Existence is that of *adequate functioning* (see further and also 1979; 2001).

We apply the term *basic* to characterize the broad, but still more limited areas of phenomena, which *together* form the fundamental area. For example, areas of phenomena of interaction of the physical particles and of the physical bodies are different *basic* areas subsumed by the fundamental area of phenomena of Motion (Interaction). Accordingly, we name *basic*—the corresponding theories and the processes. Here belong the theories of the quantum and classical mechanics. These theories and processes are well known. They represent the content of modern physics. Yabrov also is assuming a possibility of still another basic process from the area of physics—that of interaction of the micro-world and spacetime (see further).

We also differentiate the *particular* areas, theories and processes (Yabrov, 2001)—*within* the basic levels. Thus, it is necessary to emphasize that we deal with the different *scientific terms*, and not just with the mutually exchangeable semantic expressions.

In difference from the presently known fundamental theories, the new fundamental theory studies and explains Existence of *both* the *animate* and *inanimate* objects. This constitutes a defining difference of a general theory of Existence from the theories of mechanics and of evolution of species. Another principal difference of the new theory is that it studies and explains Existence of *individual* natural objects; whereas those of mechanics and evolution deal primarily with the groups of interchangeable objects. An individuality of an object and its maintenance is a cardinal characteristic of Existence and a property of an existing object (further and Yabrov, 2001). Both features—the applicability to all objects—animate and inanimate, and a necessity of consideration of their individuality—constitute the *principal* differences of the new theory from the known ones. Usually, when we generalize, we tend to equalize objects of our study. As a matter of fact, this is the main principle of generalization—finding and emphasizing the common features of things. If we cannot find a commonality, we separate phenomena into different areas of study. As a result—it became a genuine feature of the current scientific views of Nature that we separate and approach differently the animate and the inanimate objects. To overcome this deeply rooted view, we use certain non-standard key notions and approaches. It is appropriate to elaborate some of them in more details—in advance.

We speak about the "needs" of the objects applying the term to animate and inanimate objects with no distinction. This might cause some confusion leading eventually to rejection. Foreseeing this possibility, we suggest the following considerations. We deal here with a *scientific term* that has a particular meaning, which does not coincide literally with the conventional meaning of the word. By our definition, *needs* mean requirements for the maintenance of entity of an object. It is a term describing a certain tendency of the functioning of an object. "But the 'needs' presume an element of intention, which is OK for a living organism, but not for a substance, or a stone"—one might argue. This is a reasonable argument. But this is how the terms are coined. When a new phenomenon is discovered, it "needs" to be named. This could be some entirely new word, say "quark"; or a known word. In the later case, a meaning related to the phenomenon could be absent, like in terms "up", "down", "green", or "charming"—used in quantum mechanics for description of different properties of particles. Physicists invent such terms for fun—as we know, physicists are allowed to do anything. Why are they given the freedom of relaxation?—Because physicists discover the Laws of Nature (later we will see how and why it occurs). A term also may bear some initial meaning of the word, though with a definite limitation. Consider, for example, the term "stress" introduced by Hans Selye to describe both a certain influence upon an organism and a reaction to this influence (1974). A curious anecdote is related to this term. After he had introduced and published the term, Selye realized that the word "strain" would be more fitting, but the term *stress* had been acknowledged already and used by other scientists. So, Selye decided to leave it as is. This example described by Selye, illustrates that the immediate meaning of the word does not play a decisive role—important is the meaning of the term itself. We have found that the objects behave (function) in a certain way. We designated it as the functioning adequately to their *needs* (Yabrov, 1979; 1986). Further discussion should make the phenomenon and the term—clear. Yet, we should agree that an element of abstraction is necessary when we are dealing with the new phenomena and terms. Without it, we would never overcome the barrier between animate and inanimate. Closely related to the notion of needs is the use of an element of *animation* in our description and formulation of the phenomena of Existence. In our general Laws, animation speaks not about "intention", but about certain *tendency* of functioning. Simply to say—this is about function. We found that the objects exist via active functioning. This refers equally to the organisms and to the inanimate objects. Aristotle animated celestial objects—the stars and the Sun. This was not about stars being alive, but rather about them functioning. The ancients applied animation to describe active functioning when the notion and the term of function were not broadly known, yet. Physicists acknowledge that they do think of the objects of their studies in terms of animation. In the times of Bohr, a question routinely asked was: "What does atom need in order it to exist?"

Related to the above considerations is the question of the "key notions". By our definition, these notions formulate the *aspects of a phenomenon*, necessary and sufficient for its general description. A scientific study couldn't be started without discovery of the pertinent key notions (Yabrov, 2001; 2012). After these introductory considerations we come to description of the theory.

PART 7.
A GENERAL SCIENTIFIC THEORY OF EXISTENCE OF NATURAL OBJECTS AND NATURE

Conversation with the Reader

Thus, after 2500 years of continuous thought about the essence of Being, or Existence, humankind came to a simple, but profound understanding that Existence is a *physical State* of every creature, every thing, and of Nature as a whole (see above: Concept of the fundamental states of Nature). What does this insight adds to our current knowledge?

First of all, it makes the modern Man wiser and humbler. It turns out that religion and philosophy—from antiquity and to our days—are studying a problem, which constitutes the very basis of the universe and of our residence in it—Being. *Furthermore*, it becomes obvious that modern science *missed* the studying of a largest and central realm of Nature. This is a striking discovery for a scientist: an entire—pivotal—area of natural phenomena still remains unexplored. It should be acknowledged that ultimately, all the sciences study Existence (Yabrov, 2001). Physics, chemistry, biology, and other sciences bring certain knowledge about Existence of objects—using their specific approaches and methods. But this knowledge is *indirect*—rather it is a particular fraction of our overall view of Nature.

This is a staggering revelation for the lay public, which believes that science has already studied everything worthy of studying; and has explained everything (or almost so). An utterance by Smolin is appropriate here: "We are missing something big" (2006). This is the subject of persistent search from antiquity till today—by Parmenides and Aristotle, Spinoza and Kant, and in our time—by Heidegger, Wittgenstein, and Einstein.

We realize that our conclusion that a tremendous area of Nature, which constitutes its central constituent, *remains unexplained*— is a highly responsible claim. It is being presumed that what is still unclear does not exceed the general bounds of the known—the remaining problems may interest only the narrow specialists. "What more could be discovered about Nature on a general scale in addition to what the theories of mechanics and evolution have explained us?" Furthermore, "why would we even look for some additional broad explanations when everything is already understood in principle—on a general scale?"

Every age has this belief—that Nature has been understood. This state of a comparative confidence in the comprehensiveness of our current knowledge has its adaptive role. Otherwise we could not conduct our every day affairs, being incapacitated by doubts in our own power to understand what is going on around us, foresee the development of events, and act accordingly. But as an American painter Robert Smithson said: "Nature is never finished". Accordingly science, whose task is the study of Nature, is never finished either.

Structure of a theory

We have developed a general concept of a State of Existence. Our next task: to *explain How* it takes place.

The fact that things exist is obvious. To some extent, this is this obviousness which hindered a profound *scientific* study of the problem so far. By "scientific" we mean discovering the *process* underlying pertinent phenomena and the *mechanisms*, which provide for this process; as well as the regularities of the process—the Laws of Existence (Yabrov, 2001). Knowing the process, mechanisms and laws, we can explain *how* things and Nature *exist*. Answer to the How advances our understanding of Nature as a whole, and allows us exploiting new knowledge for resolution of some basic scientific problems, as well as those of our every day life, such as the problems of health and the human relationships, and of relations of Man with Nature (see this volume and also Yabrov, 1980; 2001; 2002; 2012; 2012a).

This is the *function of a concept* to encircle and describe the area of phenomena under study. The concept of a State of Existence embraces inanimate and animate objects. Thus it embraces physical reality—everything what exists. If a concept is the arch of a study, a theory is its foundation. The *function of a theory* is to *explain* phenomena embraced by the concept.

A complete scientific theory has a *certain structure*: It includes the *key notions, process, mechanisms of this process*, and its regularities—the *Laws* (see Yabrov 2001).

The Key Notions

Any new scientific theory uses the new notions (see Yabrov, 2001; 2012; 2012a). We are confronted with the new, unexplained phenomena. In order to understand how these phenomena occur, we *first of all* should study and discover the new features, or aspects that characterize these phenomena and differentiate them from those phenomena, which were known already. As a matter of fact, the peculiar characteristics, which we discover as a result of study of a novel phenomenon, these are those features that make the phenomenon *new*—different. These characteristics should be named. The new terms and meanings are not just words—they represent new concepts, or notions reflecting certain aspects of the phenomena under study. Discovery of these new notions-features is a creative process, which takes the lion share of the overall efforts needed for development of a theory. It includes observations and experiments, discovery of new phenomena, their comparative analysis, and generalizations leading to a new view. Consider, as an example, the terms "force" and "momentum" used by Newton in his descriptions of the phenomena of interaction, and in the formulations of the laws of Motion. Analogously, Darwin coined new key notions-terms—"adaptations", and "natural selection". This is why we prefer to speak of the new *terms* rather then a "new vocabulary". This is not a matter of semantics—we discover new notions.

The notions, which are *necessary and sufficient* for characterization of the phenomena under study, we name the *key notions* (Yabrov, 2001).

It is necessary to emphasize that the same key notions of Existence are valid for all existing natural objects—animate and inanimate. This amazing fact demonstrates the philosophico-scientific community of all the phenomena of Existence.

The following example illustrates a decisive cognitive role of the key notions. Imagine what it would be if we did not have the key notions of Motion, such as mass, velocity, acceleration, and others. We would not be able to generalize and to discover the physical laws. Furthermore, we could not even speak of the pertinent phenomena because we would not be able to describe them. All we could say is that things move. Precisely this situation we have now in the area of Existence. All we can say is that things exist. Having no key notions, we cannot describe, compare and analyze the pertinent phenomena. Without description and a comparative analysis we cannot generalize, and cannot formulate the laws.

Below, briefly discussed are the general key notions of Existence, i.e. those applicable for description of each and every phenomenon pertinent to this fundamental area. They represent results of generalization of our numerous observations and experiments, as well as of those performed by other investigators. We subdivide the general key notions of Existence into the following three subgroups: those *characterizing the objects; the conditions; and the functioning (or behavior) of objects* (Yabrov, 2001).

Characterization of Objects—the Notion of *Entity*

When we considered the living organisms, we emphasized a seemingly insurmountable difficulty of the study: their infinite number and diversity (Yabrov, 2001). This difficulty has increased now when we deal with both the animate and inanimate objects—consider this. Chemical substances, single-cell organisms, plants, animals, man—they are all natural objects. Yet we just listed different realms of objects. Our aim is to study how every particular *individual* object exists; this refers to any inanimate and animate objects—to a stone, an amoeba, a pine tree, a shark, an ant, a dog, and a person. Each one somehow maintains its individuality and thus exists. (Note: One could suggest an argument: Physical particles lack individuality—they are interchangeable. These kinds of the back-heel questions are suggested not so infrequently—just to probe the opponent. Our answer: the considerations of the phenomena of Existence refer primarily to the objects of *increasing complexity*. Particles exist—at least in a certain physical sense).

To understand and explain how each particular object exists, some *common* notion must be introduced allowing for a description of *every* object. This description should include characteristics of the species, and of the individual nature. At the same time, it should be applicable to any other natural object. For example, we want to understand how a particular dog—your pet—exists. We need a description that fits this dog as a species, as well as for this pet individually. This is not all. This same description should be applicable to *any* individual object, otherwise we cannot reveal the commonality, make generalizations, and specify laws valid for *all* natural objects. The task at hand is to provide a notion allowing a detailed description of an individual object, which at the same time, can be applied to any object.

We selected a limited number of basic characteristics that we consider necessary and sufficient for a description of any natural object (Yabrov, 2001). These are *composition, structure, function, individual character, experience, capacities and needs.* These characteristics are self-explanatory. They have been detailed in our analysis related to the living organisms (Yabrov, 2001; 2012a). Having these characteristics of any object, we united all of them by the notion of *entity.* It describes the attributes of the species and the individual characteristics of every object, animate or inanimate. Any object has *its* entity (Yabrov, 2001).

Dr. Alexander Yabrov

Of course, when your task is to describe features common both for animate and inanimate objects, an element of abstraction is necessary, otherwise it is hard to make generalizations. Above, we have discussed as an example the key notion of *needs*. This notion describes the requirements for the maintenance of an entity. Mindful of this definition, the reader should not have any difficulty in accepting that this notion refers both to the animate and inanimate objects. To exist as entities, the objects must satisfy needs which are determined by an object's entity.

An inanimate object is required to maintain its composition and structure in a stable form.

The primary needs of organisms are those of metabolism and self-preservation. Metabolism comprises the processing of gases, liquid, nutrients and minerals. Every organism preserves its metabolism otherwise it would cease to exist. The need of self-preservation is associated primarily with existence of animals. An individual who does not satisfy this need will perish.

The notion of entity allows for the identification and description of an object. Thus using a limited amount of certain characteristics we may differentiate and describe *any* individual object. This is achieved because each of the selected characteristics can be detailed by an unlimited amount of particular features. For example, composition covers all the possible individual chemical variations. The same refers to other characteristics of entity—each represents all the possible particular features. Entity is not just a name common for various things. This is an individualized notion.

By the introduction of the notion of entity we solved the problem of identification of any object of nature by a common notion. *Entity represents an object of nature with its species-specific and individual characteristics* (Yabrov, 2001; 2012a).

Conditions of Existence

As emphasized above, we subdivide the general key notions of Existence into three subgroups: those *characterizing the objects; the conditions; and the functioning (or behavior) of objects.*

In our interpretation, conditions describe the environment, and also the particular states of an object, which influence existence of this object. For the purposes of our study, we differentiate the *ambient* conditions of existence of objects, and *their own* conditions of existence.

Ambient conditions take place independently of an object. Here belong physical conditions, such as weather, or the day-night and seasonal alternations. We also refer to this group the general social conditions, e.g., political system of a country.

And then, there are *own* conditions of existence, which an object may influence and create. For example, by building shelters, animals create *their* physical conditions. Here we include also the *internal* conditions. Examples for the inanimate objects: solid, liquid and gaseous states of a substance; for organisms—accumulation of subcutaneous fat by a squirrel during autumn, or switch to the dormant state by some bacteria, plants and animals under unfavorable ambient conditions. As practice shows, such a differentiation helps to integrate conditions into the system of description of existence of objects. At the same time, it should be emphasized that according to the theory, the leading role in Existence play the *needs;* conditions play their role by influencing the needs (see further—the 4th Law).

Characterization of the Functioning (Behavior) of Objects: Adequate and Inadequate functioning of organisms

In what state and how long actually an individual organism exists (within one's genetic limits), essentially depends on the character of its functioning, or behavior.

Here we approach the possibility of a variation, or of *choice* of behavior. The latter refers to organisms. The more complex are the organisms, the more the possibility of choice of behavior manifests itself, and it influences the state, and actual longevity.

In order to describe the role of behavior in existence of an object, we have introduced the notions of adequate and inadequate functioning (for more details see Yabrov, 2001; 2012).

Adequate functioning (adequate function) is the behavior of an object having tendency toward a favorable state at which it exists longer.

Inadequate functioning (inadequate function) is the behavior of an object that leads to a state at which its existence shortens.

We emphasize that the notions of adequate—inadequate functioning describe primarily behavior of organisms. *Choice* of behavior is not a characteristic of the inanimate objects. Inanimate objects behave adequately. If their conditions are adequate to their needs for existence—inanimate objects may exist for an indefinitely long time. Some minerals are found whose age is equal to that of the Earth (see further—the 3rd Law).

Conclusion

The key notions allow us to describe phenomena of existence. Think of it. Until now, we knew that natural objects existed. And we could not say anything about it. We could not discuss the relevant phenomena. We could not study them, and could not explain them (and we did not).

Leading modern philosophers—Wittgenstein and Heidegger—had insurmountable difficulties being confronted by the problem of Existence (see Yabrov, 2001; 2002).   Now we have the key notions of Existence of both—animate and inanimate objects. As the following discussion shows, we do not have any difficulty whatsoever in describing pertinent phenomena. The entire area of our exploration: Existence of objects, and of Nature—is open for investigation. It is describable, knowable, and amenable for our study. We proceed with the analysis of our findings.

A New Fundamental Process

Knowledge of the known fundamental processes of physical-chemical interaction and evolution does not allow us answering the question—how objects *exist*. In order to answer the question: "How do objects exist?"—It is necessary to discover and define the *process*, which underlies phenomena of existence. A one of the difficulties of our study was that the amount of existing objects is countless, their diversity—unlimited. As described above, we have solved this problem by introducing the notion of entity. Yet, each object exists in its peculiar way. It is hard to imagine that existence of so many different objects could be studied individually, and still a *general explanation of how they Exist*—valid for every object could be discovered. To solve the problem, we need to discover a fundamental process responsible for existence of both animate and inanimate objects.

We have started from the study of Existence of the living organisms. As described above, we studied existence of the humans; these were clinical investigations of the ill and healthy individuals. Clinical studies were extended through even broader experimental investigations in various living organisms from viruses to the warm-blooded (Yabrov, 1980; 2001; 2012).

We came to resolution of the problem of the process of existence of organisms gradually. In practical terms, the general questions we tried to answer: "How an organism survives and exists? What is going on in the body under any conditions—physiological and pathogenic?"

You see a healthy child playing in the sand. An hour later, you see the same child lying in a hospital bed with high fever. Health and life of this child depend on your actions. You learned some physiological processes and their mechanisms proceeding at different levels in a healthy body. You also know how the chill develops, and how the body reacts to it. But the question is more profound and all-embracing—what is the general process responsible for existence of an organism under any conditions? Is there such a general process?

Our task was helped by the fact that the work of a physician and a cell biologist-experimenter is in fact a study of *existence* of organisms. Uncountable clinical observations and experiments with various organisms were performed by us during tenth of years (see above—Experiments). These studies led us to a conclusion that organisms satisfy their needs, which allows them to survive (Yabrov, 1979). Experiments below illustrate this conclusion.

Experiments on adaptation to needs

Below are given examples of existence of some warm-blooded organisms according to their needs under varying conditions. In these examples the attention is centered upon the adaptation and also acclimatization via training. It is well known from the studies on physiology and from our every day observations that the organisms adapt to the *changing conditions*. As a matter of fact, adaptation of an organism to the conditions is a basic principle of physiology established

Dr. Alexander Yabrov

still by Claude Bernard (1879). The founder of scientific physiology—more than a century ago—came to conclusion that organisms adapt to *conditions* of their existence. This assertion forms a basis of modern physiology and medicine. Our investigations, however, proved that the leading factor of adaptation is the *needs* of the organism. An organism adapts primarily to the needs (see further—The Fourth Law, and also Yabrov, 1979; 2001; 2012). This does not negate the importance of conditions, since the conditions may influence the needs. But needs are the *leading factor* dictating the functioning of an organism. If needs do not change, functioning remains the same in spite of the change of conditions, and vice-versa. A person, who crossed the Atlantic Ocean from New York to London at the night time, remains sleepy, though he arrives in a sunny day. And the symptoms of the jet lag persist—a reflection of functioning of the body adequately to its needs rather then to the ambient conditions, which have changed. An opposite example: an athlete is at the start of a competition—immediately prior to the start of a boxing fight. The ambient conditions—the weather, the time of the day, and the like—remain the same. But the state of the organism changes instantly—the level of stress hormones, blood pressure, and heart rhythm—all change adequately to the needs of the organism. (I have given this example because I experienced this situation repeatedly—the 2nd place in the Leningrad youth boxing championship. Nobody could say that our conclusions are not based on direct experiments and observations).

This is an achievement to find the true underlying cause of physiological reactions of organisms, which has not been known: *the satisfaction of the needs*. This discovery belongs to the area of biology-physiology-medicine and it has been discussed as such (Yabrov, 1987, 1987a). There is, however, a broader meaning in these observations, which goes beyond an area of a particular scientific discipline. We have discovered a certain *fundamental regularity* characterizing existence of organisms in general—they are *functioning adequately to needs*. Consider the following examples (from Folk, 1974; 1977; 1979).

Figure 2 shows the level of body temperature of Central American opossum—*Metachirus* at the night and day time. This animal is *nocturnally* active. According to the needs of the organism, his metabolism should be higher during the night-time; whereas at the day time, when this animal sleeps, there is no need in maintaining a high level of metabolism. This is precisely how the organism of the opossum functions. Body temperature reflects the level of metabolism.

Figure 2. Body temperature of opossum *Metachirus* at night and day time.

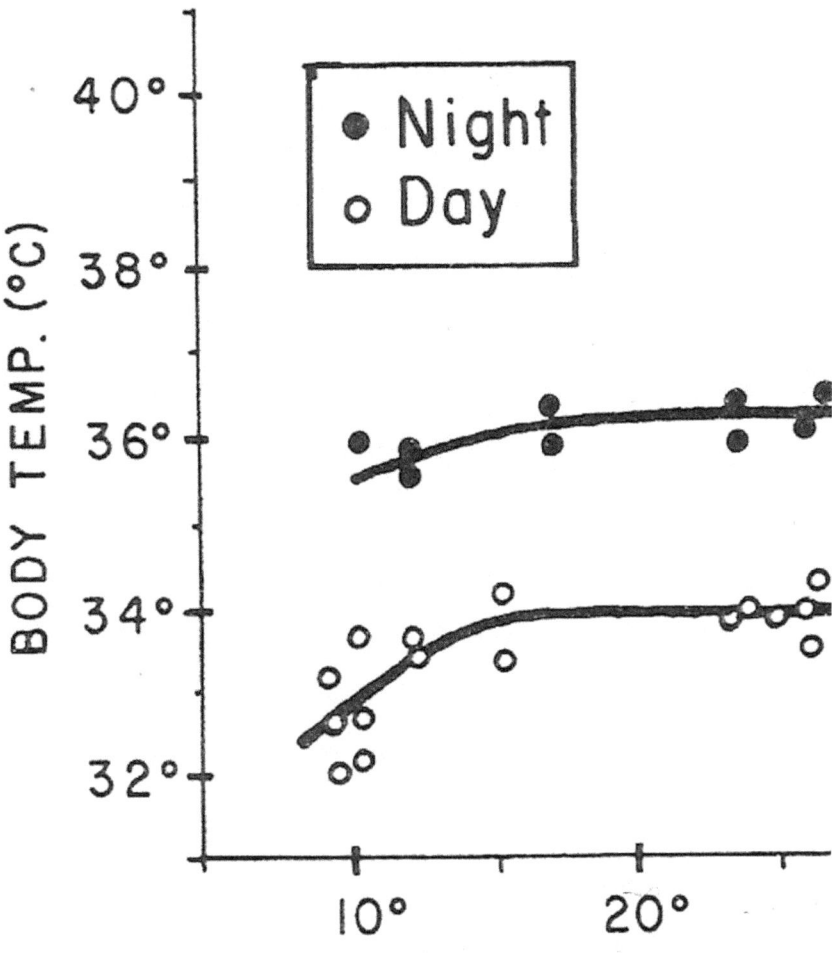

Ambient temperature (Celsius)

Dr. Alexander Yabrov

It is seen that the *body* temperature of opossum is increased at night and diminished during the day at the same ambient temperature.

The next example illustrates the role of a preliminary exposure to the adverse ambient conditions (acclimatization) upon the functioning of the organism. Needs of a trained person differ from those of the untrained individual. This is reflected in their reaction to the same conditions of an experiment (Fig. 3).

Figure 3. Heat production of men before and after cold exposure.

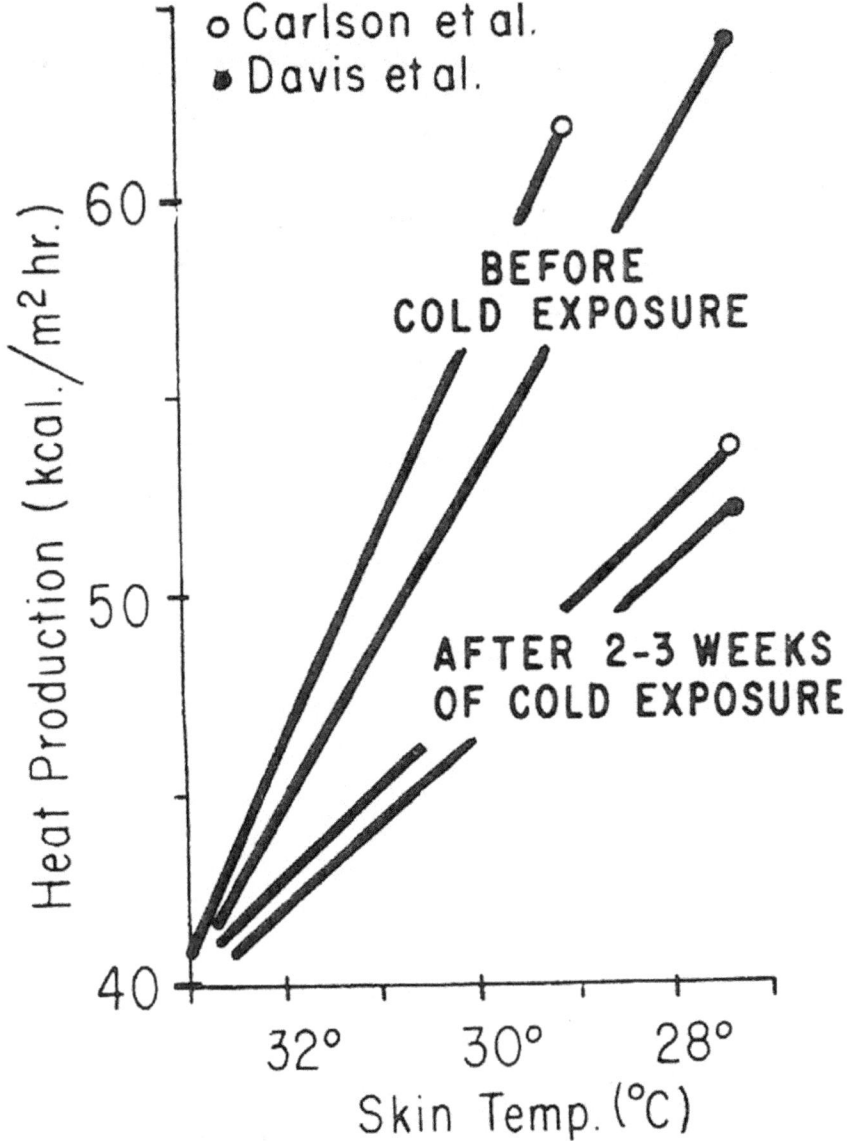

Cold-acclimated subjects had a lower heat production than controls in a cold test. The graphs above are remarkable because the results were so similar when obtained independently in two different laboratories.

Figure 4 illustrates difference in the ability of individuals with different training to adjust to cold.

Figure 4. Metabolism and skin temperature during sleep in cold.

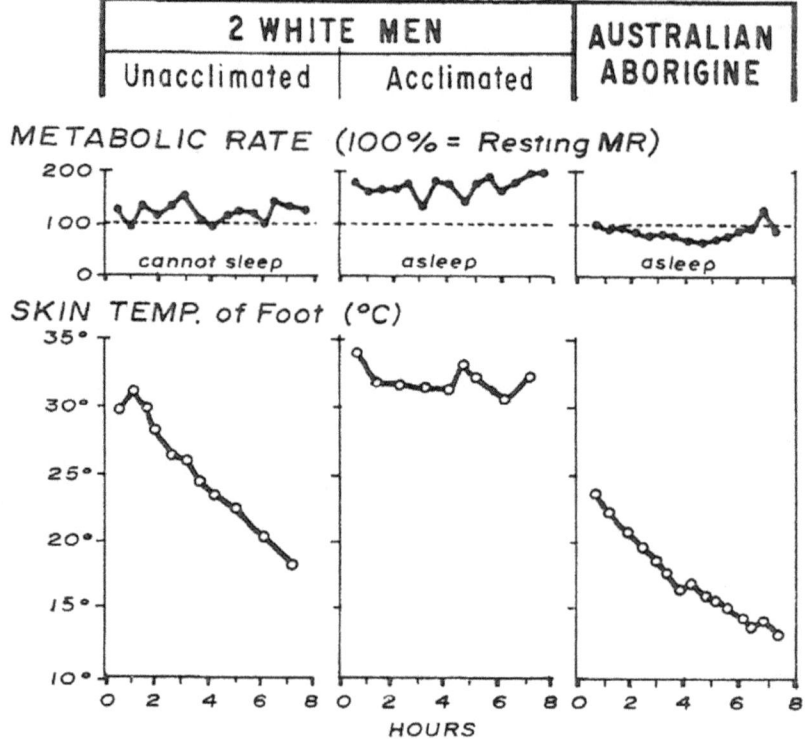

Australian aborigine is accustomed to sleeping in cold. Needs of his organism do not change essentially. His metabolic rate is stable and a bit lower then at the day time, which is physiologic for the state of rest. His skin temperature is diminished as a way of preservation of body energy. White men are not accustomed to sleeping in cold. This is a potentially pathogenic condition for them. Functioning of an acclimated individual reflects mobilization of energy metabolism adequately to increased needs. The level of metabolic rate is enhanced;

it is somewhat unstable; skin temperature is increased. Still he is able to sleep. The unacclimated is unable to function according to the challenging needs. He cannot sleep. His metabolic rate fluctuates at a day time level; skin temperature after a slight spike fells down precipitously. This person may become ill. The experiment reflects adequate functioning in accordance to difference in needs.

The following example illustrates the changes of functioning of an organism according to needs under high humidity (Fig. 5).

Dr. Alexander Yabrov

Figure 5. Exercise and acclimatization to humid heat.

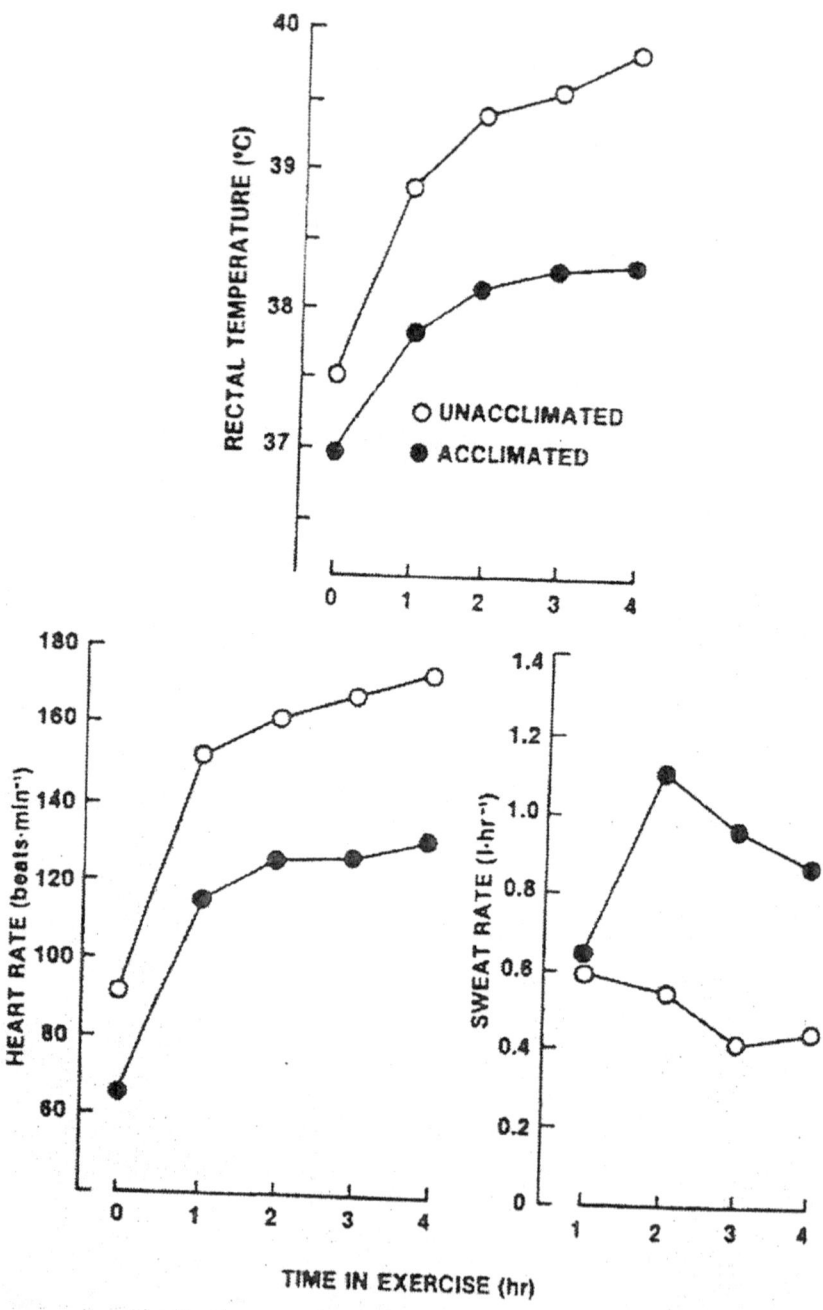

TIME IN EXERCISE (hr)

In Figure 5, the graphs compare responses of male subjects during 4 hours of exercise in humid heat, before and after a program of acclimatization. Experimental conditions are the same, but the organism's needs have changed and the functioning changed adequately.

Among the mechanisms allowing an organism to survive at a very low ambient temperature is shivering. The function of shivering is to increase heat production by adding the work of muscles. Because the work-function of this muscle contraction is zero, shivering happens to be a very economical thermo generator: hit production increases 3 to 4-fold. The onset of shivering has been used as a very effective test of ability to resist cooling. Studies were performed in Korean diving women—Ama—who are cold-acclimated as a result of many years of training. These subjects experience the most severe form of cold stress that humans voluntarily tolerate for a long time. E. Folk (1997) describes the procedure as follows. They dive as deep as 20 meters, wearing a ballast belt (2 kilograms), and holding on to another heavy weight of up to 15 kilograms. The diver holds on to this weight and sinks quickly with it. As soon as the work is done, the boatman pulls her quickly to the surface (Figure 6).

Dr. Alexander Yabrov

Figure 6. Gathering of natural corals from the sea bottom.

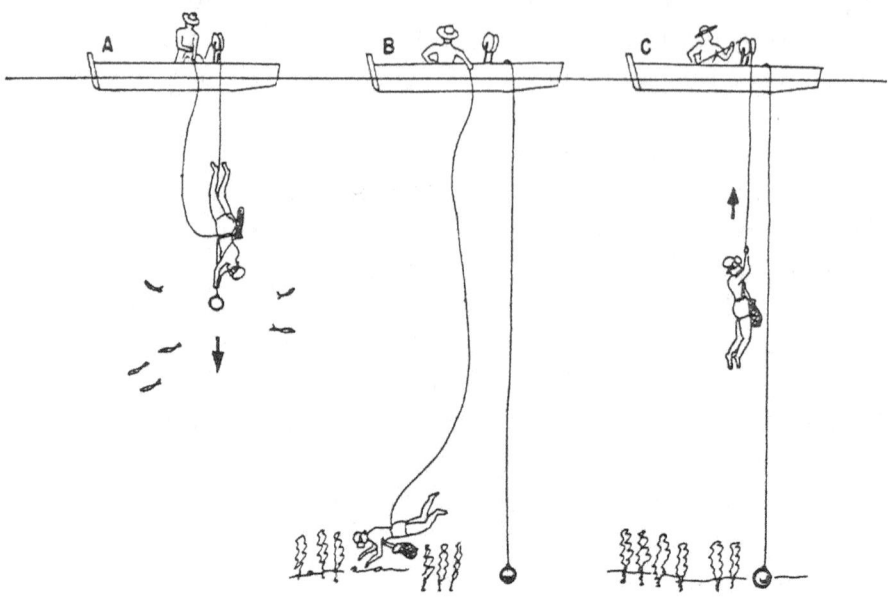

Diving women of Korea and Japan (Ama) work in the deepest cold water, descending rapidly by holding onto a weight, and later ascending rapidly with the help of an assistant who pulls their life rope up over a pulley.

Physiological studies showed that Ama are more resistant to cold (Figure 7).

Figure 7. Shivering threshold of Ama.

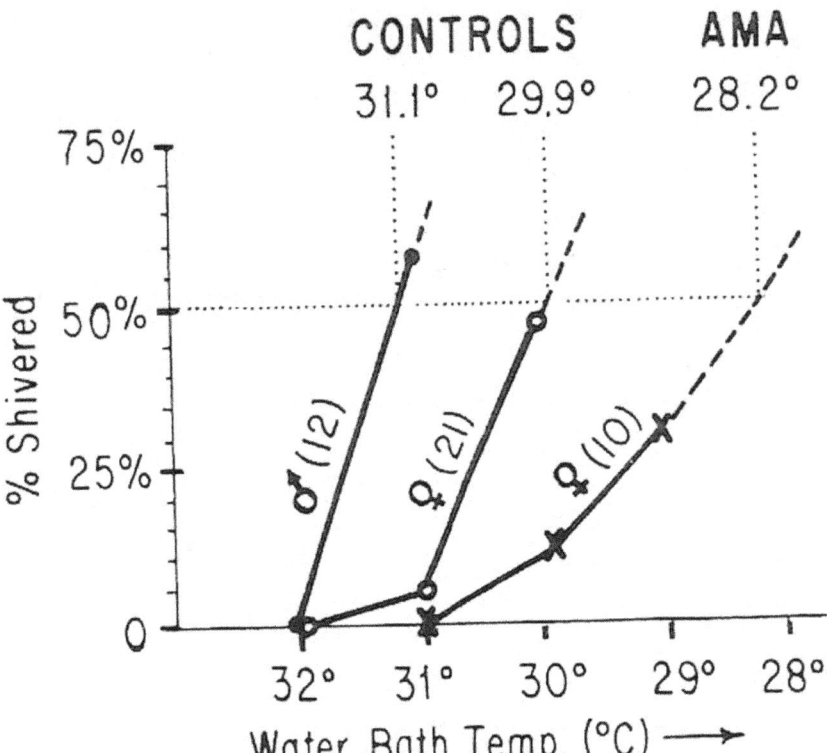

Figure 7 shows that Ama became cold–acclimatized. Therefore they need an enhanced level of hit production only at comparatively lower ambient temperature. Adequately to needs, the average shivering threshold of Ama is lower (by temperature) than that of the non-divers—both women and men.

The functioning adequately to needs of the organism is not limited by the examples of adaptation. Consider the following simple example. If you pour out a drop of chlorine acid into a Petri dish with amebas, these single-cell organisms will swim away; if you pour a drop of food—amebas will swim toward it. This is an obvious functioning adequately to the needs of the self–preservation and the maintenance of metabolism, correspondingly.

Dr. Alexander Yabrov

The process of Adequate Functioning

The examples presented above are sufficient for the reader to discern certain regularity. Analysis of the results of the much broader studies of functioning of organisms eventually led us to a simple generalized conclusion: organisms are functioning adequately to needs directed to their survival (Yabrov, 1979). Based on our further investigations, however, we have reformulated this principle: *organisms function adequately to needs directed toward the maintenance of their entity* (Yabrov, 1986; 2001). Note: this formulation allowed us to apply the law of adequate functioning to both animate and inanimate objects (see the 1st Law).

Try to answer the following complex question: "What each human individual is doing throughout the entire life?" Our answer based on long-term studies is: "Every person functions adequately to the needs directed toward the maintenance of entity". This, of course, includes all actions aimed at the survival. But broader than that—this definition covers also the actions directed to satisfaction of various needs of a person, such as social, creative, sexual, and any others related to the maintenance of entity of this individual. The principle of adequate functioning is valid for any organism.

Adequate functioning—a new term

We have realized that discovered the process of existence of organisms (N). It was clear to that it needed a particular name. After long deliberations we have coined a new term—"process of adequate functioning" (or "adequate function").

Adequate function of an organism can be defined as the use of one's capacities, i.e., properties, internal resources, mechanisms and experience adequately to needs directed toward maintenance of entity under various conditions—physiological or pathogenic (for more details see Yabrov, 1979; 1987; 2001).

As a matter of fact, the task of a physician is to help the organism in its adequate functioning under the conditions, in which it may find itself. Judicious use of our understanding of the concept of adequate function based on the knowledge of the general trend of reactions, and of particular mechanisms responsible for them allows us approaching the solution of the major task of contemporary medicine—prevention and treatment of chronic diseases (see Yabrov, 1980; 1987; 1987a; 2001; 2001a).

Since *functioning* of an organism represents the leading subject of our discussion, the difference of the meanings of the terms "process" and "mechanism" needs to be clarified. Colleagues advised us to choose one of these terms throughout the entire analysis of phenomena—*or* the process, *or* the mechanism. The point is, however, that this is not a matter of semantics. We are dealing with two *different* essential constituents of a phenomenon. A phenomenon cannot be understood without both discovery and consideration of each of these components, and, of course, a phenomenon is impossible without them. In order to illustrate the difference between a process and a mechanism, which provides for the process, we put forward an example where a mechanism is hand-made; it is literally a tool. The flight of an airplane is a process. The plain's motor is the tool, or the mechanism, which provides for the flight. It is clear from this example that a process (flight) and the mechanism (motor) are different things.

Definitions. A *process* is a sequence of changes or steps that determine the occurrence and character of phenomena. For example, inflammation is a particular process acting at the cellular-tissue level. *Mechanisms* are the natural means that make the process possible. The mechanisms of the process of inflammation are certain cellular and humoral factors (leukocytes, antibodies, interferon, and others).

In order to describe the *mechanisms* responsible for the process of adequate functioning, we need to consider the structural-functional organization of organisms (see further—the Second Law).

Encouragement by the colleagues—oxygen for a discoverer (a personal experience)

I realized that I have discovered new general principles of Biology: Existence of organisms and the process, which governs it. At that time (1977-1978) I was working in a hospital laboratory, in Canada—a newcomer—having no creative contacts or an audience—to report and to get a critical opinion. I needed an authoritative profound evaluation. I find it appropriate to describe again what I have mentioned above about my search for creative contacts. My experience might be useful for some of the readers.

I have summarized my results and presented the compendium to Sir Karl Popper, a world renowned English philosopher-scientist in the areas of Biology and Physics; and to James Watson—Nobel Laureate, co-discoverer of structure of the DNA (USA). Soon I have gotten the answers—both highly positive. Especially close creative interactions, which lasted for many years, has developed with Sir Karl. He immediately recommended my materials to the "Medical Hypotheses". In January 1979, I got a letter from the editor:

"I have just received a letter from Sir Karl Popper. He believes that you are an outstanding scientist and has asked me to help by publishing some of your work in my journal, Medical Hypotheses. I am very willing to do this and I enclose a sample copy of the Journal. I look forward to hearing from you. With best wishes, yours sincerely, Dr. David Horrobin."

An author—any author—could not get a better letter from an editor. Without delay, I have sent an article—"Maintenance of adequate function is a general principle of survival of organisms". It described a *new area of biology*—Existence of organisms, and *how it occurred*: via the process of adequate functioning. Professor Horrobin invited me to visit his laboratory in Montreal. I arrived by bus—I did not have enough money to fly from Princeton, where I started working on production of interferon for treatment of cancer. "Your embrace is broad"—David told me, approvingly. We went through the entire article word by word—the editor wonted to preclude any possibility of misreading caused by my English, though he did not emphasize it to me. Could you find an editor of a scientific Journal, who would allot that much of attention to a paper by an unknown author? The article—"Maintenance of adequate function is a general principle of survival of organisms"—was published in three months—in 1979.

Next year—another article has been published—"Adequate function of the cell: interaction between the needs of the cell and the needs of the organism". It provided ground for a new view of physiology and pathology. The same year, my book "Interferon and Nonspecific Resistance" has been published in New York. It was awarded as the Best Book of the Year 1980 in Health Area, by the American Publishers Association. It included definition of adequate functioning.

The revelation

An amazing and stimulating feature of science is that you did not know something very important, you even have not thought about it (at least consciously)—and suddenly—minutes later—you know it—as if you knew it always. See below.

I came to the general idea of Existence and of a new view of Nature through my studies in Medicine and Biology.

Says Einstein:

"It has happened often in physics that an essential advance was achieved by carrying out a consistent analogy between apparently unrelated phenomena. In these pages we have often seen how ideas created in one branch of science were afterwards successfully applied to another" (Einstein, Infeld, p. 270; 1960).

This is precisely what has occurred in the course of our studies of Existence of
organisms.

Needless to say that all our investigations were performed with the aim to help a physician in his work. This principle is reflected in our publications: about the growing problem of chronic diseases (Yabrov, 1985; 1986a), rehabilitation of the impaired brain function (Kanevsky, Yabrov, 1987), mechanism of pathogenesis and treatment of AIDS (Yabrov, 1986; 1988; 2000), and others. These studies led to discovery of a general mechanism of chronic diseases (1987a), and eventually to the new theories of biology, physiology, pathology, and medicine (Yabrov, 1979; 2001a, b; 2012; 2012b; Yabrov, Okunev, 2004).

It is impossible, however, to limit your contemplation by a sole field of natural phenomena even when it is that broad as the life phenomena are. Understanding of the process and the mechanisms of existence of organisms had served us as a *base* for a *broader* comprehension of Nature. The ideas started glimmering in mind—that the adequate functioning was somewhat a broader happening—not limited by the life phenomena. Then, some day—this probably was spring of 1984—a revelation came.

A serious scientific revelation is, of course, an unforgettable event for the researcher. Years, or as in our case—decades, of persistent experimental probing of facts—guided by a certain idea—suddenly brings you to an unexpected generalization: A generalization, which you have not foreseen, and which redirects your entire study. For every investigator a revelation comes in its peculiar way. A Russian mathematician, who found a solution to a very difficult theorem, described that in the middle of the night when he was sleeping, Peter the Great visited him and dictated the precise formulation. What remained on his part was—to wake up and to write down the formula. (Needless to say that this happened in SantPeterburg—the city founded by Peter the Great).

In our case, it happened in the day time—in Princeton. (Princeton is a place where ideas are growing on the trees—what remains—is to pick them, I believe). I was not sleeping, but probably I was not completely awake either. I was roaming like a somnambulist in the study room; some unorganized, subconscious thought hovering in my mind without any certain idea or direction. Suddenly—the reality of the day has burst into my conscience. I started realizing where I was; I clearly have heard the voices in the neighboring room and the intrusive TV talk. I began seeing sharply—as if somebody put glasses on my nose. An idea struck like a lightening: "Existence embraces everything—the *inanimate* objects *also* exist by the process of adequate functioning".

Immediately, I started the work in this entirely new direction: development of a *general* theory of existence of natural objects—not only the living organisms, but of *both*—the animate and the inanimate objects—via adequate functioning. Colleagues and friends advised that I should have stopped this deviation from my "really important work"—in the area of biology and medicine. But how could you close your mind and refuse to hear the call of the *unknown*? Nature has opened slightly the veil covering a *new area* of phenomena—never studied as such by modern science—the fundamental area of Existence. All my previous studies—clinical and experimental—were in fact, a spade-work for the discovery and study of Existence in its general embrace.

A general principle of existence of natural objects

But what does this adequate functioning of the inanimate objects preserves and maintains? Not the *survival*, of course. It maintains the being of an object as a certain individuality—an entity. Preservation of the individuality of an object as such during an appreciable period of time is what Existence of this object means. Note: Speaking of "natural objects", we have in mind primarily the objects of increasing complexity (see further—definitions of Existence).

As I have mentioned, after the conclusion that both the living organisms and the inanimate objects exist via the process of adequate functioning, I have sent an article to the Medical Hypotheses: *General principle of existence of natural objects*. Its Abstract said the following:

"Natural objects, live and inanimate, actively maintain their entities by using their capacities (i.e., properties, internal resources, mechanisms, and experience) adequately to their needs. It is suggested that the principle of adequate function is a law of nature" (Yabrov, 1986).

After the editor-in-chief, Dr. David Horrobin, has read the manuscript, he remarked in his letter—as follows.

"History knows examples of the medical doctors discovering the laws of Nature. Since the article claims a discovery of a new Law of Nature, we have invited three outside reviewers in addition to our three internal reviewers. Instead of three months, the review should take at least six months."

In about half a year, another letter from the editor came. It said that the overall opinion of reviewers was positive. But by the opinion of one of them—something similar was described by Lovelock in his "Gaia hypothesis". Therefore it was necessary to analyze that theory and find the differences, if any.

Gaia is the name for Mother Earth used by the ancient Greeks. Lovelock and Margulis (1974) define Gaia as a complex entity involving the earth's biosphere, atmosphere, oceans, and soil. According to the hypothesis, in order for life on earth to exist, certain conditions on its surface are to be preserved within the limited favorable range.

Dr. Alexander Yabrov

These conditions include temperature, concentration of particular gases, humidity, water salinity and pH, and some other physical and chemical parameters. As Lovelock stresses, geological records show that these conditions were actually maintained during the period of history of life on earth despite the profound changes in the environment (Lovelock, 1979; 1988). For example, the earth now receives 1.4 to 3.3 times more energy than it did just after its formation. Still the temperature of the earth's surface is kept within the life-supporting average range of Io to 30°C for hundreds of millions of years from the time when life began (Lovelock, 1979). The Gaia hypothesis postulates that the physical and chemical condition of the surface of the earth, of the atmosphere and of the oceans has been and is actively made fit and comfortable by the presence of life itself (Lovelock, Margulis, 1974). In our opinion, Gaia can be considered as a *living complex entity functioning adequately to its needs* (Yabrov, 1986). Besides its heuristic role, the Gaia hypothesis has practical importance for the understanding of the problem of pollution (Lovelock, 1988). Lovelock mentions that the start of the Gaia hypothesis was the view of the Earth from space, revealing the planet as a whole but not in detail (Lovelock, 1979; 1988). Ecology—he says—is rooted in down-to-earth natural history and the detailed study of habitats and ecosystems without taking-in the whole picture. The one cannot see the trees in the woods. The other cannot see the woods for the trees (Lovelock, 1988).

Gaia hypothesis represents an *illustration* to our theory of adequate functioning. Its description was inserted in our article with an explanation that adequate functioning covered the woods and the trees. It covered the cells composing the trees and the one-cell organisms. It also covered the inanimate objects. The article was published (Yabrov, 1986).

Our Reasoning for the General Principle of Existence

Natural objects, both live and inanimate, actively maintain their entities; otherwise they would not exist. To maintain the entity for an organism is to be alive. For a substance it is to retain its composition in some stable form.

In order to exist as entities, the objects have certain needs to be satisfied. These needs are determined by the nature of an object.

For an animal, such needs are metabolism and self-preservation. An inanimate object needs to maintain its composition and structure.

The objects satisfy their needs using their properties, internal resources, mechanisms, and experience, which we named 'capacities' (1986; 2001).

Needs are influenced by the conditions of existence. Conditions may influence the possibility of satisfying the needs. There are conditions at which organisms cannot exist at all. For example, life of an earth-borne organism would be impossible at 200°C, or being immersed in a strong acid or alkali.

There are certain physical and chemical environmental conditions under which organisms may satisfy their needs directed to the maintenance of their entities as living beings. The character of these conditions and their limits depend on the nature of an organism.

For example, temperature limits, light of a certain wavelength, gaseous, or aquatic media, all these conditions of existence are different for a human, a fish, a plant, or a bacterium. The limits and the optimums of these conditions further vary to some extent for each individual organism.

The same can be said about the substances needed for metabolism and of their quantities necessary for existence of various organisms.

For example, for a human these are water, oxygen, complex nutrients, and minerals. Their quantities are different for different humans depending on their size, sex, age, and character of metabolism.

For a plant, or a bacterium, the simple substances may be sufficient for the satisfaction of their needs.

Variations in conditions within the limits of existence of a given organism may influence the character and level of his needs.

For example, under some steady conditions, an animal has a certain level of metabolism. When the conditions change (e.g., increased physical loads, or cold, or starvation), a transient process of changing of needs takes place in the organism resulting in increase (or decrease) of the level of metabolism in accordance with the new needs. The new level of metabolism is then maintained unless the conditions change again. Other examples of the influence of changing conditions upon the level of metabolism are hibernation and also spore formation.

Under any conditions the organism tends to function adequately to his needs determined by his nature, and directed to the maintenance of his entity.

One may conclude that live organisms maintain their entities under various conditions using their capacities adequately to their needs determined by their nature.

Inanimate objects also exist by maintaining their entities adequately to their needs.

Similar to what takes place with live objects, the needs of inanimate objects are determined by their nature and influenced by conditions.

Inanimate objects do not have needs of metabolism or self-preservation as animals do. Still, substances need to maintain their composition and structure determined by their nature.

Similar to what takes place with live organisms, there are limits of physicochemical conditions at which different inanimate objects may exist as entities. For example, sodium chloride is very stable under usual conditions. Its content in sea water reaches 3%. Meanwhile, sodium cannot exist as such under usual conditions. Because of its high reactivity, sodium should be stored in the atmosphere of an inert gas, or be immersed in mineral oil.

So, we have around us only those objects, which can retain their entities under the usual environmental conditions.

Variation in conditions within the limits of existence, e.g., of a substance, may influence the state, or the form in which it exists. For example, each substance may be in a state of a gas, a liquid, or in a solid state under different conditions.

It is determined by the nature of the substance—in which state a given substance exists under the given conditions. In all these states the substance functions adequately to its needs directed upon the maintenance of its entity, i.e., of its composition in a stable form under the given conditions.

Sometimes a substance in a one and the same state, e.g., crystal, may exist in two or more stable forms. Properties of the same substance in different forms differ. Compare, e.g. diamond and graphite, which are different crystals of carbon.

However, in both these forms of existence the internal forces—atomic, molecular and intermolecular—are directed toward the maintenance of carbon as a certain substance. Thus, one may say that in both forms the capacities of the substance are used adequately to its need to maintain its entity.

Probably the most demonstrative example of an adequate function of an inanimate object is represented by the solar system.

For the purpose of our discussion, we assume that mass of the planets is the only property, and gravitation is the only force responsible for the maintenance of existence of the solar system under constant conditions. (Note: The example of solar system was suggested by Professor Slonim of the Beer-Sheva University, Israel).

In order for the solar system to exist as a stable system, i.e., in order that its entity is retained, it is necessary that the planets composing it do not fall upon the sun, and do not flow away from it beyond the force of gravitation. The only way that these needs can be satisfied is that the planets would continuously move around the sun by certain orbits determined by their mass. This is exactly what is observed. This means that the solar system maintains its entity by functioning adequately to its needs. In this example, motion and gravitation play role of the mechanisms of existence.

The solar system is composed of several natural objects whose interaction forms a qualitatively new complex entity. Being in itself a natural object, it exists in accordance with the principle of adequate function.

Analogous conformity with the above principle is traced when we analyze the existence of complex animate entities, e.g., those formed by the organisms in a biocenosis. In the most general form, this is illustrated by the Gaia hypothesis (see above): Gaia exists via adequate functioning.

Dr. Alexander Yabrov

The new view of nature allows us to *unify* phenomena of inanimate and animate worlds.

Gap between inanimate and animate worlds represents a seemingly permanent insurmountable disunity in Nature. I suggest, however, that this depends on our *dominant view of Nature*. Currently, we view behavior of *physical bodies* from the view of Motion—the process responsible for Motion is that of *physical-chemical interaction*. And we consider behavior of *living organisms* from the view of evolution. Thus we have two areas of objects and phenomena—inanimate and animate—considered from *different views of Nature* underlain by different fundamental processes. Their unification is *impossible*. However the situation changes if we consider *all* natural objects and phenomena from a different point of view—that of *Existence*. Both—inanimate and animate objects *exist*. As we discovered, the fundamental process responsible for their existence is that of adequate functioning (Yabrov, 1979; 1986). It should be emphasized that so far there was no a common principle, which would describe various individual natural objects—live and inanimate. This is the principle of adequate functioning. We thus came to the broadest generalization of existence of natural objects. It unifies the inanimate and animate worlds of objects and phenomena. This unification is a result of a change of our view of the World. An entirely new panorama of Nature opens to us. The new view brings profound understanding of the harmony of Nature. A harmony, which Einstein aspired to discover and explain. We see an *Existing United World* of the natural objects and phenomena inanimate and animate.

PART 8.
GENERAL LAWS OF EXISTENCE OF INDIVIDUAL NATURAL OBJECTS

What is a Law of Nature?

We enter now the temple of science—The Laws of Nature. The reader might think that a law of nature is a well defined notion—otherwise, how can we speak of laws, and rely upon them, if it is not even clear what constitutes a law of nature? Yet the question remains an open one. Doubts exist as whether there are the laws of Nature—at all.

This collision of views is illustrated by the following statements dating to different eras.

At the close of the sixteenth century modern science discovered the first laws, and at the onset of the next century Galileo wrote:

"Nature acts through laws which she never transgresses".

In the twentieth century, Wittgenstein pronounced:

"The whole modern conception of the world is founded on the illusion that the so-called laws of nature are the explanations of natural phenomena. Thus, people today stop at the laws of nature, treating them as something inviolable...It is hypothesis that the sun will rise tomorrow: and this means that we do not *know* whether it will rise." (Wittgenstein, pp. 140, 142; 1961)

Popper's view closely resembles Wittgenstein"s:

"Why do we succeed with our theory-making? Answer: We have succeeded so far, and may fail tomorrow...And the fact that horse-drawn vehicles could be seen every day in London for many centuries has not prevented their disappearance and replacement by the motor car. The apparent "uniformity of nature" is quite unreliable; and although we can say that the laws of nature do not change, this is dangerously close to saying that there are in our world some abstract connections, which do not change...and that we call them 'laws of nature". (pp. 98-99, 1972)

<end excerpting>

Those, who do agree that the laws of nature exist, cannot suggest a definition. Of the criteria considered so far are—*predictability* and *necessitation* (Armstrong, 1983). The above example by Popper of the carriages replaced by the cars demonstrates unreliability of the former of these criteria. Schrödinger's definition is the following:

"Now, what we call a 'law of nature' is nothing else than any one of the regularities observed in natural occurrences, insofar as it is looked upon as necessary..." (Schrödinger, pp. 135-136, 1935).

Our arguments against this definition: If any regularity were a law of Nature, we would have an uncountable amount of the latter, which is not the case; not every regularity which we observe, could be qualified as a Law of Nature. Necessitation is a subjective and therefore unreliable criterion. Also, we do not have here a definite criterion for differentiation between a law and a coincidence.

An opinion shared by many is that regularity in relations or a series of phenomena that hold together can be considered to be a law based on fact. As to which one can be considered a law of Nature, this is "...explained in terms of people's attitudes" (Ayer, p. 234; 1973). For example, the law of gravitation is viewed as a law of nature, while boiling of water at 100 degrees (Celsius) at standard pressure is considered too specific to be classified as a law of nature. That such an important scientific problem as the law of nature should be left to somebody's psychological judgment is—in my view—unacceptable.

Our explanations

Definition

Thorough physical analysis led us to the following conclusions. Law is a description of regularity of a pertinent *process*. The process assures stability of a law. So far as the process of interaction exists, we can be sure that the sun will rise tomorrow. (The problem what is a Law of Nature discussed in more details in Yabrov, 2001).

What differentiates a law from a coincidence?

Difference between a *stable regularity* of the interrelations among the pertinent phenomena, which we name a law, and—a coincidence, is determined by the involvement (or its absence) of an underlying *common process*. The community of the process assures persistence of regularity. A certain infectious disease is caused by a certain microorganism (not by a different one, and not by cold, or miasma)—a law assured by the process of infection. When I visit New York, it rains—a coincidence; phenomena of raining and of my visiting are not underlain by a common process (Yabrov, 2001).

What law does qualify to be called a Law of Nature, and what does not?

The subject under consideration is important for development of basic science; yet it is not sufficiently elucidated.

We do not consider a psychological judgment to be an appropriate answer to the above question. A decisive role plays the *character of the underlying process*. If you have doubts, whether your discovery deserves being qualified as a law of nature—try to *characterize* the common process underlying phenomena under study: Is the process fundamental, basic, or particular?

We differentiate the fundamental, basic and the particular processes (above, and Yabrov, 2001). By our definition, *laws of nature describe regularities of interrelations of phenomena immediately underlain by a fundamental, or a basic process.* Therefore these laws describe the regularities observed within a one of the entire *fundamental* realms—those of Existence, Motion or Origin, or at least, within a one of its *basic* levels.

A regularity that is immediately underlain by a *particular* process—unites a limited circle of phenomena. Such regularities *are* laws (and not coincidences), but usually they are *not* qualified as the laws of nature. Rather, they constitute the laws of a certain scientific discipline. For example, laws of infectious diseases are related to medicine; laws of motion of the planets—to astronomy. Yet, we should emphasize that some regularities underlain by a particular process, which unifies a broad circle of natural phenomena of high importance for Man, besides being the laws of particular specialties, are also classified as the laws of nature. Here belong, for example, the laws of electricity.

It becomes clear why physicists are so successful in discovering the laws of Nature, as well as the broad laws of the scientific disciplines, such as those of electricity, light, etc. Physics studies the regularities of phenomena *immediately* underlain by the fundamental, or by the basic processes of interaction, or by the broad particular processes from the area of Motion. Other sciences, in contrast, study only the phenomena immediately underlain by some narrow *particular* processes. For example, medicine studies the course of different diseases and the methods of their treatment. The processes immediately involved are so particular that the regularities discovered cannot be qualified as the laws of Nature. Research by both physics and medicine is important, since it promotes amelioration of the human condition approximating a favorable state. The laws described below are immediately underlain by the *fundamental* process of adequate functioning valid both for the animate and inanimate objects—therefore they are qualified as the Laws of Nature—General Laws of Existence.

Just above, we have defined the meaning of the notion of a Law of Nature: *A regularity of a common fundamental, or basic process.* This definition allows answering some unresolved problems of physics.

On the problem of disunity of physical laws

In his lecture, *The Character of Physical Law,* Feynman states:

"No one has ever succeeded in making electricity and gravity different aspects of the same thing. Today our theories of physics, the laws of physics, are a multitude of different parts and pieces that do not fit together very well." (p. 29, 1990)

Feynman develops his idea further:

"We do not have one structure from which all is deduced; we have several pieces that do not quite fit exactly yet. That is the reason why in these lectures instead of having the ability to tell you what the law of physics is, I have to talk about the things that are common to the various laws; we do not understand the connection between them" (Feynman, p. 30, 1990).

I suggest that the differences may be explained by that the theories of physics and the laws they describe—consider phenomena underlain by *different processes*; "no one has succeeded in making electricity and gravity aspects of the same thing", because the regularities, or laws of electricity and those of gravity *immediately* describe the regularities of *different basic* processes of interaction—those of particles, and of physical bodies: thus, they "do not fit together very well." This is how we explain the *disunity* in physical laws.   Another   topic touched upon by Feynman in the same lecture is that of the *scientific predictions*:

"How is it possible that we can extend our laws into regions we are not sure about? Why are we so confident that, because we have checked the energy conservation here, when we get a new phenomenon we can say it has to satisfy the law of conservation of energy?"

And Feynman adds:

"If you will never say that a law is true in a region where you have not already looked, you do not know anything. If the only laws that you find are those, which you have just finished observing, then you can never make any predictions. Yet the only utility of science is to go on and to try to make guesses. So what we always do is to stick our necks out, and in the case of energy the most likely thing is that it is conserved in other places" (Feynman, p. 66, 1990)

Feynman is absolutely right—the generality of the law of conservation considered above adds certainty to our guess. The *commu-*

nity of a process is the central issue. We may predict the validity of a law based upon consideration of the process involved. If the process that underlies the phenomena, which has not been examined before by our study, is the *same* as the one immediately described by our law, we may expect it will be valid in the realm, which we are about to explore.

The *new* laws are needed to be discovered

A law of Nature cannot embrace and explain phenomena not immediately underlain by the process responsible for the regularity described by this law. The same refers to the *theories* describing the laws of nature. Physical laws and the theories explaining these laws describe phenomena underlain and proceeding through the fundamental process of physical-chemical interactions. But there are natural phenomena underlain and proceeding by *different* fundamental processes and by the corresponding basic, as well particular, ones. To describe and understand these phenomena, *other laws* and *other theories* must be discovered.

Currently, science considers all natural phenomena in a sole framework—within the mechanistic (or—mechanistic-evolutionary) worldview. Our unified picture of the world is based on a well substantiated understanding that the same physical laws are valid throughout the Universe. Physical laws are considered to be the *only* Laws of Nature. However, the following utterance by Popper provokes serious doubts about the comprehensiveness of the today scientific depiction of the world by the physical laws.

"Almost all regions of the universe are filled by chaotic radiation and almost all the rest by matter in a similar chaotic state....If the picture of the world which modern science draws comes anywhere near to the truth...then the conditions obtained almost everywhere in the universe make the discovery of laws of the kind we are seeking...almost impossible." (Popper, 1985)

Obviously, Popper does not consider our knowledge of physical laws to be sufficient for our complete understanding of Nature. What kind of the laws is Popper talking about? I suggest that these are the laws that should help us to understand the phenomena taking place under the conditions of existence of the stabilized complex objects. These are the conditions of our everyday existence. These new laws—"of the kind we are seeking"- should allow us understand and explain the phenomena, which we routinely observe and of which we are the immediate participants. We come to a definite conclusion that different laws of nature are needed to be discovered in addition to those known already. These new laws are described below.

PART 9.
THE LAWS OF
EXISTENCE

General laws are valid for both animate and inanimate objects

Phenomena of Existence remain unexplained by the laws of Motion and Origin because phenomena of Existence are underlain by a fundamental process that is *different* from those of interaction and evolution. As described above—this is the fundamental process of *adequate functioning* (see also Yabrov, 1979; 1986; 2001; 2002).

It should be acknowledged that the laws of existence were initially drawn by us with the aim at description of existence of *living organisms* (Yabrov, 2001). All previous attempts by other investigators to discover the laws of life sciences have failed (Keller, 2003). We have formulated these laws so that they correspond and do *not contradict* the description of existence of the *inanimate* objects.

Thus, there is a difference in the prevailing explanatory orientation between the laws describing phenomena of Motion, and those describing phenomena of Existence. The objects of consideration of the physical laws are primarily the inanimate objects, though the laws of physics do not contradict the life phenomena; and to a certain extent, the physical laws are applicable for understanding of the live Nature. In difference from the physical ones, the laws of Existence are primarily aimed at description and explanation of existence of living organisms, but they do not contradict and are applicable for understanding of existence of the inanimate objects. The explanatory overlapping of the laws of Nature—those of Motion and of Existence—is a reflection of the *unity* of Nature. Our success in a broad

Dr. Alexander Yabrov

understanding of Nature depends on our unified view of natural phenomena. For example, as we will see, physical laws are applicable for the explanation of existence of inanimate objects since Motion is a one of the mechanisms of Existence (See Table 1).

We use some methods helping the "bridging"; we hope they will not be rejected for the semantic reasons. An element of animation should not confuse the reader. We should not allow our analysis and description of Nature to be thwarted by some traditional taboos. Our goal is to understand and explain scientifically—how the natural objects exist—the task *never tried* before. We apply "animation" to describe some phenomena of the inanimate nature in order to describe certain *functional* manifestations. There might be a confusion concerning the laws holding an element of animation—implying an idea of intention, or purpose to the description of functioning of both the animate *and* inanimate objects. In essence, however, we convey the idea of a certain *tendency,* or direction of the adequate functioning toward the maintenance of existence. This way we emphasize the essence of adequate functioning, which is confirmed by our experiments and the broad observations of natural phenomena. It cannot be excluded that some new key notions will be discovered in the future for the description of the pertinent aspects of the phenomena of Existence. Revealing the regularities of their interrelations might result in discovery of other laws. Existence is a new field of a scientific study and it should be expected that some new notions and the laws would be developed in this area. Also, it cannot be excluded that some laws introduced by us might prove not corresponding to characterization as the laws of nature, or to be unnecessary at all.

The laws of existence are presented below. They formulate the regularities of interrelations among the general key notions of Existence. The phenomena described by these laws are immediately underlain by the fundamental process of adequate functioning. Therefore these laws qualify as Laws of Nature: the General Laws of Existence of natural objects (Table 1).

Note: Description of the *general* laws of existence, which follows below, has much in common with the description of the laws of existence of organisms (Yabrov, 2001; 2012a). This should have been expected. Yet, we have decided not to abbreviate their presentation. Analogy of description emphasizes the analogy of the Laws of Existence of animate and inanimate objects.

TABLE 1. A Brief Account of the General Laws of Existence
1. Law of the maintenance of Existence of objects.
2. Law of the structural-functional organization of objects.
3. Law of adequate and inadequate functioning.
4. Law of adjustment to needs.
5. In Nature, objects Exist only in-groups.
6. Law of interaction of objects in-groups.

We also have discovered some *particular* laws of certain areas of natural phenomena, which still could be classified as the laws of nature in the area of existence. Here belong, for example, the certainty constant (Yabrov, 2012a), and a law of adequate social behavior of Man (Yabrov, 2001; 2002; 2012).

A Review of the General Laws

Six laws are presented. All laws are valid for both animate and inanimate objects. Laws of Existence can be subdivided into two groups.

The first group—the invariant laws, consists of two laws.

The First Law describes how every individual object exists, and discovers that all objects exist through active functioning; this refers in full measure not only to animate, but to inanimate objects as well. The law explains many phenomena of existence and also allows us to answer in a general form as how individual objects exist.

The Second Law describes how all natural objects are organized. It uncovers regularity in the *organization* that is *common* to *all* existing objects notwithstanding their obvious diversity. In other words, no matter how different objects may look, a definite community exists in their organization. It concerns *structures,* which are similar in different objects, and also the *processes* and *mechanisms* relating to the structures. As will be seen, many phenomena of existence become understandable based upon the Second Law.

Dr. Alexander Yabrov

The second group comprises four laws, which like those of the first one are valid for all natural objects. Yet, a certain peculiarity differentiates these two groups of the laws of Existence.

The first two laws describe regularities, which are *invariable*. All objects exist actively maintaining their entity otherwise they cease to exist (1st law). All objects are organized according to a structural-functional organization common to all individual objects. All objects include an atomic level of organization. The following organizational levels are also common for the objects of analogous complexity (invariable law). This is the organizational foundation of natural objects, animate and inanimate (2nd law).

The laws of the second group, however, stipulate a definite *variability* in the way that the active function of existence is fulfilled by an object. The functioning may bring an object to a favorable state and maintain it in this condition, thereby allowing this object to exist potentially longer—*adequate* functioning. Or the process of existence may be implemented in such a way that it prevents or hinders the attainment of favorable state thereby cutting short existence—*inadequate* functioning.

At first glance, admitting the possibility of variation in the ways of fulfillment of the laws of nature may appear unfeasible. The view—that the way in which a law of Nature manifests itself—is fixed—follows from knowledge about the laws of mechanics: For example, the Earth revolving around the Sun (invariable law). Another example: the reader knows that if he jumps up, no matter how high, he *will* return to Earth. This is how the law of gravitation manifests itself (invariable law).

Laws of existence differ from those of classical mechanics by allowing variability of behavior of objects. They differ also from the laws of quantum mechanics by that the variations of existence are not the results of a chance determined by a statistical probability. Variations in behavior of the objects based on the laws of existence are the manifestations of adequate functioning. The laws of Existence describe regularities, which take place in Nature. If these laws did not describe the variations, they would not correspond to reality. Consider the example of animals. Individual organisms have a choice in behavior: The more highly organized an organism is, the more diverse its behavior—man is at the maximum end of the scale, a single-cell organism—at the minimal one. Inanimate objects do not possess choice. Variations occur when conditions change—thus changing the needs of existence of an object (e.g. change of temperature, pressure, and the like).

To be able to characterize behavior of every object and, at the same time, to have an opportunity to arrive at the general conclusions valid for different objects, we consider how the behavior of an object influences its longevity—a quantitative attribute of existence. This is how we came to the notions of adequate and inadequate functioning (Yabrov, 1969; 1979; 2001). A classification based upon these two notions—those of adequate and inadequate functioning—permits characterization of any behavioral variations of any object. The Third Law states that the way objects exist may vary, thereby influencing the duration of existence.

The Fourth Law describes how objects function depending on their needs under varying conditions. The Fifth and Sixth laws describe the interrelationships and interdependence of the existing objects (see also Yabrov, 2001; 2002). Below follows an elaborate discussion of the laws of Existence.

Detailed description and analysis of the laws

Having defined what constitutes a law of Nature (a regularity of underlying fundamental process) and presented the laws of existence in general terms, we may discuss now each law in more details.

The First Law: Law of the Maintenance of Existence of Objects

Objects of nature exist by satisfying needs directed toward the maintenance of entity.

This law was formulated about twenty fife years ago (Yabrov, 1986). It explains in a general way *how objects exist*. The law states that objects exist through active functioning, but the description does not stop there. It specifies this function as the one of satisfaction of needs directed to the maintenance of the entity.

Entity is a notion that describes the individual *and* the species-specific characteristics—i.e. the features of *every* natural object. We identify an object by its entity. This an unusual hybrid of meanings is encapsulated in a single notion: a meaning that retains the uniqueness of each object and—at the same time—a meaning encompassing any object. With a two-faceted notion, we may make a generalization of how objects exist that is valid for every natural object. Objects exist by maintaining their entities. This is achieved by satisfaction of needs, namely those directed to preserving the entity. Thus by explaining how the objects Exist, we answer Leibniz's question—"Why is there something rather than nothing?" And—furthermore, we answer—in principle—the fundamental question posed by Wheeler: "How come existence?" (see further).

Needs may be various. As a key notion, this term designates the requirements necessary to preserve an entity—of both animate and inanimate objects. Needs directed to the maintenance of a stone's entity are those directed at preserving chemical composition and structure, including its unique size and shape, for example. Until these needs are satisfied, the stone exists as an entity.

Similar considerations are valid for any plant. To exist as an entity, it needs to keep its composition, structure, appearance and metabolism. A pine tree, for example, is a complex object. In addition to a stone's atomic-molecular and the body levels, the tree also possesses cellular and organ-system levels. The needs are different at each level: at the atomic-molecular level—this is to maintain an interaction among the atoms and molecules. At the body level: keeping the physical object as a whole. At the cellular level—it is to maintain the cellular metabolism. At the organ-system level it is to maintain interaction among the organs and systems so that the needs of the organism as a whole are satisfied.

Now we can explain how humans exist (N). Look at what people are doing every day for the duration of their lives. Each individual is a unique entity engaged in different day-to-day activities. It may seem impossible to put forward a general characterization of all the diverse actions of individuals. The First Law provides for the answer to the question—"What is each person doing?"—It pertains to everyone: every individual functions in accordance to needs directed toward the maintenance of ones entity. In a general form, this refers to any natural object.

The Second Law: Law of the Structure and Function of Objects

Natural objects exist according to a common structural-functional organization.

The Second Law describes two features characterizing existence of natural object.

Both are central for the description and understanding of Existence.

Firstly, the Law says that all natural objects represent a *unity of structure and function*. There are no objects represented by the structure alone, as there are no objects represented only by the function; or by anything else, for that matter.

Secondly, the Law says that there is a certain organization in how all objects are built. This organization concerns both the structure and the function, and is *common* for all objects. We know that many objects are complex and endlessly different. The 2nd Law reveals an *order* in this complexity. The difference is in appearance, whereas the basic organization of objects is certain and it is *common* for all objects. In other words, objects are *not different* as far as it concerns their organization—and this is a Law.

We define a natural object as the one that occupies space (Yabrov, 2001). We do not consider an idea as a natural object. This can be an atom or a molecule, a grain of sand or a solar system, an amoeba or a human. When we analyze all these different objects, inanimate and animate, we always uncover certain structural-functional levels of existence *common* to *all natural objects*.

Why should these levels be taken into consideration? Because our goal is not only to know *what* exists, but also to understand *how* objects exist. Aristotle's *Great Chain of Being* comprised minerals, plants, animals, man, angels, and God. The Philosopher thus introduced an idea of nature's organizational **complicacy** by presenting a sequence of objects of increasing complexity. This was a description of *what* exists.

Dr. Alexander Yabrov

Our aim is of a different order: It is to provide an explanation of *how* things exist. Accordingly, the Nature's organizational pattern is presented here in a different way. Instead of considering particular existing objects as such, the world of things and events is classified by the structural levels and their corresponding processes and mechanisms responsible for the innumerable phenomena of Existence pertinent to each of the levels. These levels can be found in each object depending on its complexity. For example, a chemical substance is represented by one level—the atomic-molecular, whereas a complex social animal—man—represents a whole hierarchy of the levels of existence.

An investigator of existence of individual objects should not ignore the picture of the united world of phenomena and, at the same time, should not lose oneself in its complexity. Consideration of the law of the structural-function organization of existing objects serves these purposes. The law designates the organizational mode of existence of objects, firstly, by the levels of their increasing structural complexity.

It is indeed remarkable that all individual natural objects, which appear so different, have an organized structure *common* to all objects. This is one of the facts (or truths) that the Second Law discovers. It describes a limited number of *basic* structural levels, each level being the *same* in every object. An object's complexity is determined by the amount of levels. The community of the structural organization of natural objects is the first regularity, which this law uncovers and describes. Furthermore, it discovers that the *structural* community is accompanied by the *functional* one.

Functional aspect is represented at each level by the *basic processes* and the *mechanisms* which together are responsible for the *overall fundamental* process of adequate functioning. In contradistinction from a fundamental process, which is *common* to all organizational levels, the *basic* processes and mechanisms of Existence *differ* by the *levels* (Table 2).

TABLE 2. Basic Structural-Functional Levels of Organization of Existing Objects (modified from Yabrov, 2001).

Level	Process	Mechanisms	Scope of phenomena
Atomic-molecular	Interaction of particles	Motion, elctro-magnetism, strong and weak forces	Ubiquitous
Bodybuilding	Passage from the state of uncertainty into that of certainty	Electromagnetism, inner gravitation, surface energy, spacetime grip (?)	Frontier of the micro-world
Physical bodies	Interaction of bodies	Motion, gravitation	Phenomena related to physical bodies
Cellular-organ-system	Metabolism	Staple, regulatory, compensatory, defense	Living organisms
Organism-behavioral	Habitation	Reflexes, instincts	The area of interaction with the environment
Social (general)	Social relationships	Comprehension, experience, way of behavior	The area of interaction with others
Social (human)	Human relationships	Reason, morals, law	Interactions in the human society

This Table represents a development of a similar one described earlier (Yabrov, 2001). The atomic-molecular level includes the sub-level of Bodybuilding. It describes interaction of the micro-world and the macro-world, and its mechanisms—inner gravitation, surface energy, and the spacetime grip—the latter ones were discovered by us after the publication of *How Man Exists* (see for details Yabrov, 2012).

Table 2 is rather simple; nevertheless it describes how each and every individual object—inanimate or animate—exists. It presents the basic levels of organization of the existing objects, basic process-es and mechanisms acting at each of these levels, and the scope of phenomena belonging to every level. It should be emphasized that the Table describes the *basic* levels. Each of them comprises differ-ent *particular* structural-functional levels discussed elsewhere; for example, the basic *cellular* level comprises many particular levels of organization—those of metabolism of the nucleic acids, proteins,

Dr. Alexander Yabrov

carbohydrates and lipids; energy metabolism, and others. Each characterized with its structures, processes and the mechanisms (see Yabrov, 1980a, b; 1987; 2001). As Table illustrates, a definite common order of organization is characteristic for all the existing individual objects—despite of their seeming unlikeness. The first two levels are common both for the inanimate and the animate objects. In addition to these, organism has also other levels. At the cellular-organ-system level, we differentiate two groups of mechanisms. The first includes the *staple* mechanisms—those responsible for existence of an organism as a living unit. Here belong the mechanisms, which provide for the genetic and metabolic functions. Another group consists of the *regulatory, compensatory and defense* mechanisms. These are the mechanisms of adjustment, which provide for the maintenance of functioning of staple mechanisms adequately to the changing needs (Yabrov, 2001; 2012).

As seen from the Table 2, if we compare the mechanisms of the *same levels*—they are the same for all objects. For example, at the atomic-molecular and the body levels of organization of a stone, or of a man—the same mechanisms—motion and the forces of interaction—are acting. If we compare *different* organizational levels, we would not find the similarity of the mechanisms, but this would be an erroneous comparison. This is the essence of the principle of structural-functional organization that the mechanisms belonging to the same organizational levels are the same, but at the different levels the mechanisms are different (see for more examples Yabrov, 2001).

This differentiating role of the mechanisms of different levels should necessarily be taken into account when generalizations are made. For example, it is erroneous—I repeat it again and again—to ascribe to the quantum mechanics the ability to explain all natural phenomena. The point is that the mechanisms acting at the atomic-molecular level and at the other levels of the structural-functional organization of natural objects are different. Mechanisms acting at different levels cannot replace each other. Quantum mechanics explains how physical particles move and interact, but it cannot explain how physical bodies move and interact, or how amebas, or people function. Anderson's thesis: "More is different" (1972)—is valid precisely because mechanisms acting at various levels of the structural-functional organization are different.

It is hard to come to any generalizations when dealing with a gathering of an uncountable amount of dissimilar individual objects. The principle of structural-functional organization presents us the world of existing things in a form amenable for a general analysis. Knowledge of the basic (and particular) processes and mechanisms allows us to understand what kinds of phenomena are taking place at each level, and how they occur. It is seen from the Table that every organizational level plays its special role in existence of an object; none could be omitted or replaced by the other level. The processes-mechanisms of every level fulfill their certain function. If a structural damage or malfunction occurs at any of the levels, this impairs the function—i.e. the processes and mechanisms acting at this level—and thus distorts existence of the object as a whole. For example, if a person has stomach ulcer, the whole body suffers because function of the local cellular mechanisms is impaired. Acting together, the basic processes and mechanisms of different levels provide for the general fundamental process of adequate functioning that governs existence of the individual objects of various complexities (see the 1st Law).

If we compare the basic structural levels of Existence with those of Motion and Origin, a certain feature becomes evident. Motion and Origin do not take an individual object into consideration—they consider the groups of objects. For example, in the area of Motion, the structural level of physical bodies unites all visible bodies (macro-world) into one group, and the level of invisible physical particles (micro-world) into another.

Similarly, one of the structural levels of the Origin of terrestrial objects includes all geoformations (inanimate world), the other—all the species of organisms (living world). These classifications are appropriate because they allow discovery of the corresponding processes and their mechanisms acting in these areas (Yabrov, 2001). When we describe phenomena of Existence, however, our consideration is centered upon *individual* objects. We must remember that an individual natural object is *the* object of our study. The aim of this study is to discover how every object exists. Accordingly, the Second

Law discovers the structural organization of *individual* objects: without knowledge of it we cannot proceed with our investigation of existence of objects.

The middle column of Table 2 describes the basic mechanisms providing for existence of objects at each basic structural level. The Table shows that the subdivision into certain structural levels of organization is determined by mechanisms acting at a corresponding level. This emphasizes the structural-*functional* nature of organizational levels. For example, the atomic and molecular structural organizations are presented as a unified level because of the community of the basic mechanisms providing for existence at both these levels. Whereas the organism-behavioral and social levels, although they refer to the interaction of the organism with the environment, are considered to be different ones, since different mechanisms are responsible for existence at these levels.

It should be emphasized that the fundamental process of an object's existence as a whole is provided by the *concerted* action of *all* the mechanisms. But in the Table 2 the mechanisms are presented strictly in accordance with their levels. For example, the forces of interaction are mentioned only when the atomic-molecular level is considered, although these mechanisms are part of the existence of every object, animate and inanimate—as a whole. This is what it means when we say that objects exist as a result of all the mechanisms acting in concert, although the mechanisms acting at the atomic-molecular level are not shown at other levels. The same refers to other basic processes and mechanisms shown at their corresponding levels. This way of presentation avoids unneeded redundancy in the Tables. It also allows the reader to differentiate the mechanisms according to the levels of their action.

Description of the processes and mechanisms is mostly self-explanatory. We elaborate briefly the description of the social level.

Social, or *social-behavioral* is the highest structural-functional level of organization of organisms. Social behavior primarily implies interaction with other similar organisms.

We name *comprehension, experience,* and the *manner of behavior* among the mechanisms of social interaction. These are *general* descriptions of certain species-specific mechanisms regarding social situations. To interact with the others, an organism should comprehend the social aspect involved. In the course of interacting, an organism uses one's social experience and behaves in a certain way. The above refers to organisms of various complexities: for example to various bacteria, or insects, such as ants or bees, as well as to the warm-blooded animals, like, for example, elephants. In each case, however, the *means*, or mechanisms used for social interactions vary. This is determined by the species-specific and individual capacities of the organisms. Thus, in difference from all other organisms, mechanisms of adjustment *of a human* at the social level are Reason, Morals and the Law. This is a great advantage of the theory of Existence that it included the latter mechanisms in its framework. So far, the contribution of basic science in understanding of the human social relationships has been close to nil, or negative. *Morals represent a foreign notion for mechanics and evolution.* There is nothing moral (immoral) in the interaction of particles or in the random genetic mutations. Yet adequate functioning of humans is impossible without Morals, as well as Reason, and the Law (Yabrov, 2002; 2012).

Hierarchy of the levels of Existence

Consideration of the 2nd Law reveals a certain *hierarchical principle*: *behavior* of an object is determined by the processes-mechanisms of a *higher* level of organization. For example a stone behaves according to the laws of classical mechanics, though it comprises also the atomic-molecular level (for more examples see Yabrov, 2001).

The Principle of Appropriateness

It follows from the 2nd Law that a particular or a basic scientific theory should be based upon an *appropriate fundamental theory* since the latter discovers the process underlying *all* the pertinent phenomena. For example a theory of human relationships should not be based upon the evolutionary theory (Marx's mistake). It should be based upon the theory of Existence. We name this correspondence of the theories—the principle of appropriateness. Disregard of the latter inevitably leads to erroneous conclusions and recommendations (Yabrov, 2001; 2002; 2012a).

Dr. Alexander Yabrov

Note: Some skeptic might argue that the author speaks of the known things: it is well known that objects are somehow organized. Well, this is a feature of the laws of nature that they speak of phenomena observed by everyone. But the Laws reveal the properties not previously recognized and thus enrich our knowledge—the 2nd Law is an illustration. Not only has it said that the objects are composed of the structural-functional levels of increasing complexity. These levels are *similar* in *all* objects, as are the processes-mechanisms acting at each level. Moreover, those of the higher level determine *behavior* of the entire object. The words by Schrödinger fit here: "The task is not to see what has never been seen before, but to think what has never been thought before about what you see everyday."

It is more productive for the reader to pay attention to the novelty, than to cherish one's skeptical mood by rejecting it. The 2nd Law has a broad application (see Yabrov, 2001).

The Third Law: Law of Adequate and Inadequate Functioning
In the process of their existence, natural objects are functioning adequately or inadequately

The laws of classical mechanics do not assume that an object's behavior may vary under constant conditions. But the phenomena of Existence cannot be described and understood without taking behavioral variability of the objects into account. The notion of existence is closely connected to the notion of longevity, which depends on the state of an object, whereas the state depends on conditions, as well as on an object's behavior.

To characterize behavior, we introduced the notions of *adequate* and *inadequate functioning*: The former results in a favorable state and greater longevity. The latter hinders the achievement of a favorable state and shortens existence.

Just as the previous laws, the third law refers both to animate and inanimate objects. Inanimate objects, in most cases, behave adequately. Existence of the solar system demonstrates an adequate function of a complex inanimate object (Yabrov, 2001). In order its entity be maintained, the planets comprising it must not fall upon the sun or fly away. The only way these needs can be satisfied is for the planets to continuously revolve around the sun. This is precisely what is taking place. This example illustrates the laws of Existence. The solar system exists by satisfying needs directed at the maintenance of its entity (the first law), it represents a certain structural-functional organization (the second law); and it is functioning adequately—to remain in a favorable state where its longevity is assured (the third law).

We might speak of behavioral *variation* of inanimate objects in the process of their existence under changing ambient conditions— e.g. transition from a solid state to liquid, and then to gaseous—under changing temperatures; but cannot speak of a behavioral *choice*.

Living organisms can exercise *choice* in their behavior; observations show that in an overwhelming number of cases they function adequately using their capacities. A rabbit seeing a man may choose to run away, or to stay motionless. In both cases this rabbit functions adequately—to avoid the danger and thus to promote ones existence in a favorable state.

Manifestations of functioning relating to existence may be complex, for example, as in the case of the interaction of a predator and its prey: their needs are conflicting. Analysis shows that both function adequately, so as to maintain their entity in a favorable state.

It becomes especially difficult to trace the functional trend when the existence of complex animate entities is analyzed; for example, those formed by organisms in a biocenosis—a term signifying a community of diverse organisms existing together under natural conditions. Nevertheless, it is possible to uncover adequate functioning. In the most general form, this is illustrated by the Gaia theory (Lovelock, 1979; 1988).

The Gaia theory postulates that the physical and chemical conditions of the earth's surface, of the atmosphere, and of oceans have always been made habitable by the presence of life itself. Consider Lovelock's examples. Sunshine and warmth promote the growth of vegetation. Forests maintain the atmospheric gas composition and humidity necessary for the living creatures to exist in a favorable state and to prevent soil erosion. Clouds provide protection from excessive radiation and heat. Rains provide for the redistribution of water bringing it to the areas where it is necessary for maintenance of life. Based on these facts, Gaia can be considered as a complex entity functioning adequately (Yabrov, 1986). It is appropriate to emphasize here that the human activities damaging the natural medium should be qualified as inadequate functioning on the part of the humans.

It is possible to find examples when individual organisms do not strive to maintain their entity in a favorable state. By their own actions they shorten the duration of their existence. This is what we call *inadequate* functioning. These examples can also be explained by the theory of existence.

An example that immediately comes to mind is that of a whale swimming ashore: findings suggest that animals displaying this kind of behavior are ill. It is possible that their sensory organ responsible for the animal's sense of orientation is damaged. The whale might be functioning in accordance to some needs (e.g., hunger). Yet this activity is a manifestation of inadequate functioning since it prevents the maintenance of entity in a favorable state that results in shortening of existence of an object.

The following example may be put forward. Curiosity broadens knowledge and leads to the acquisition of useful experience. However, in some cases, it might be a manifestation of inadequate functioning. The alertness of a curious squirrel may be so distracted by the movements of the fox's tail that it misses the opportunity to escape from this versatile hunter. Other examples of an inadequate functioning of animals can be given (see Yabrov, 2001).

Examples drawn from human behavior explain why inadequate functioning is considered to be a manifestation of existence. One can take the case of a person who commits suicide. It may be argued that this is an extreme example. While this is so, all the preparatory actions leading to a suicide are manifestations of existence. The major aim of this behavior is to shorten existence—of one's own and (or)— of others. Hence we qualify it as "inadequate".

Note: These days we see an increasing amount of the suicidal bombings. This behavior is aimed at destruction of many lives. It is definitely an inadequate behavior. In this situation the 3rd Law fulfills a vital social role. Our differentiation of the adequate and inadequate kinds of behavior provides a *criterion for an objective evaluation of hu-*

man behavior. Now no-one may suggest that the notion of a terrorist is "relative": "A terrorist from your point of view—is a freedom fighter from somebody else's view." An act aimed at the shortening of life of the innocent people is a manifestation of inadequate functioning. It should be condemned and prevented for the sake of Existence of humanity.

Suicidal behavior is a straightforward example of inadequate behavior. In a daily life, the characterization of behavior is far more difficult. The theory explains that a comprehensive favorable state of an object can be achieved only when functioning at *every level* of its organization is adequate. The laws of adequate functioning, of maintenance of entity, and of structural-functional organization—interact. A drug addict might believe that there nothing wrong with his habit. He satisfies one's need of "getting high". It is a feature of his entity. Is he acting adequately? Unfortunately, intoxication of cells by a narcotic gradually makes them unable to fulfill vital needs at the cellular-organism level. Health of this person declines. Since the brain cells are involved in this pathological process, his personality changes, this leads to a distortion of his social functioning. He disregards the needs of others; they, in turn are less inclined to consider his. The person becomes isolated, cutoff from the mainstream. His entity degrades, his life shortens.

The examples discussed just above are those showing a conscious choice of inadequate behavior. Inadequate behavior might also be the consequence of ignorance, lack of experience, or the wrong assessment of a situation.

Example. In France (and in other countries) birds like to perch upon the warm rails of the railway line. When they feel the vibration of an oncoming train they fly off in proper time, which is a manifestation of adequate functioning. However, when in France the superspeed trains started to operate, birds perished in droves because they did not expect the train to appear that rapidly after they initially felt the vibration. This is a manifestation of inadequate functioning as a consequence of lack of experience.

It is possible that a person's behavior hinders the attainment of a favorable state and shortens existence as a result of a deviation from adequate functioning not evident to this individual. A person may not opt for the best choice of behavior in a certain situation. Furthermore, an individual may not even know that his behavior is inadequate. For example, the fact that smoking is injurious to health was proven scientifically only recently. Or take the scientific recommendations on nutrition as another example. Previously we were told that tomatoes were hazardous to health, because they promoted development of gout. Today tomatoes can be eaten, but not meat. Controversies of this kind are as manifold as the number of advisers. Because these recommendations are not individualized, they are never optimal for each person. Another factor complicating a choice is the continuously changing conditions of existence. Therefore continuous favorable state is the one, which an individual tends to, but never actually achieves. Since eventually the favorable state is manifested by the maximum duration of existence, this notion is as hypothetical as that of giving a precise age for the maximum human life span.

But the notions of adequate and inadequate functioning are *not* hypothetical for they describe how a natural object exists. Observations show that life presents examples of both adequate and inadequate functioning. Behavioral variations are endless. An analysis of these observations shows that an individual's state and longevity depends on prevailing behavioral patterns (within genetic limits).

These conclusions result in the following two Articles of the third law:

Article1. *Phenomena of existence present a spectrum of adequate and inadequate functioning*

Article 2. *When adequate functioning prevails, objects come closer to their favorable state and exist longer.*

The above formulations might seem dry, but subsequently, when the theory of adequate functioning will be applied in practice for understanding of man's existence as well as that of other objects at different organizational levels, it should become obvious that the law of adequate functioning including the above Articles, provides an explanation of the multifarious manifestations of the phenomena of Existence.

Now we will discuss the role of needs and conditions.

The Fourth Law: <u>Law of Adjustment to Needs</u>

When ambient conditions change, natural objects functioning adequately alter their conditions of existence using their capacities and experience so that their needs are fulfilled

Above, in the discussion of the key notions, two kinds of conditions of existence of objects—the *ambient* conditions, and *their own* conditions of existence—were differentiated. Ambient conditions are the surrounding general ones that take place independently of the object, such as the whether. The own conditions are those under which an object exists immediately, for example, conditions of the dwelling. Here also belong the conditions of the state of the object itself. For example, an animal can be in a state of activity, or hibernation. An inanimate object can be in a solid state, or in a liquid one, according to the needs under changing ambient conditions.

This is a feature of a scientific method that conditions are necessarily considered when certain regularities in Nature are described. This is reflected in such terms as "under given conditions", "under the observed conditions," or "all the conditions remaining unchanged".

Physics' great achievement was that it included consideration of conditions in its methodology. Scientific experiments are performed under carefully controlled conditions. This allows comparison of experiments performed at different times and by different investigators—leading to reproducible and reliable results.

The theory of natural selection recognizes the role of conditions in the origin of species.

From the time of Claude Bernard and Louis Pasteur, the investigations of the individual living objects by physiology and medicine also attached importance to the conditions under which observations were made. This approach marked the start of the era of scientific physiology and medicine in the second half of the nineteenth century.

From the point of view of Existence of objects, however, it is necessary to take into consideration also the role of *needs*. After all, this is *the possibility of fulfillment of the needs* of maintenance of entity which determines the fitness to the conditions.

Dr. Alexander Yabrov

Different inanimate objects can exist as entities under a limited variety of physical-chemical conditions. For example, sodium chloride is very stable under general ambient conditions therefore its content in sea water reaches 3%. Meanwhile, sodium cannot exist as such under these conditions. Because of its high activity, sodium should be stored in the atmosphere of an inert gas. Thus we are surrounded by those objects, which retain their entities under general environmental conditions. This also applies to living creatures: there are conditions under which the organisms cannot exist at all. For example, existence of an earth-borne organism could not be maintained at 200°C, nor when immersed in a strong acid or alkali.

Analysis led me to conclusion that the suitableness of conditions is determined by a possibility of *fulfillment of needs* for the maintenance of entity.

Under certain physical and chemical environmental conditions organisms may satisfy their needs directed at maintaining their entities as living beings. The character of these conditions and their limitations depend on the *needs* determined by the organism's entity.

For example, temperature limitations, light of a certain wavelength, gaseous or aquatic mediums—all these ambient conditions of existence differ for a human, a fish, a plant, or a bacterium in accordance with the needs of these species. The limitations and the optimum of these conditions vary to some extent also according to the individual needs of each organism. And here we inevitably come to consideration of the role of needs in behavior and existence of a natural object.

As mentioned above, there are ambient conditions under which an object cannot exist; this occurs because the *needs* of the objects cannot be satisfied. Within the boundaries of conditions under which an object exists, it functions so that its needs be fulfilled. An organism functions according to one's needs. When needs remain constant, despite a change in ambient conditions—organism functions according to its needs: a person, who arrived in Europe by air in the morning, is sleepy, in spite on the sunny day. An example of an opposite situation: needs may vary while conditions remain constant. Subsequently, the function changes according to needs, such as those of an insect that goes through different stages of metamorphosis in a very brief period of time when conditions do not change.

One can imagine a group of people existing under similar conditions. In a real life situation the conditions of many individuals are similar in many cases. The behavior of every individual under similar conditions corresponds to his "needs". Each person acts differently from others when needs differ, or individuals may act in concert if they have a common goal (needs).

This does not mean that conditions are not important for existence, on the contrary, their role is essential as they influence needs and the possibility of their satisfaction; the latter, in turn, has an immediate influence upon the state of an object. For example, in the desert, an organism needs water to assuage his thirst. If this need is not sufficiently satisfied, eventually the vitality of the organism declines and the duration of existence shorten.

Thus the needs of existence and the conditions are tightly interwoven. We have examples when change of the ambient conditions causes change of the state of an object, or of the needs, or both. For example, when the ambient temperature goes down, the temperature of a substance changes accordingly, the distances between molecules and atoms shrink, their fluctuations become more constrained and slower. When the ambient temperature increases, the changes of an opposite character take place. As a result, at different certain temperatures a substance may turn into a gas, into a liquid, or become solid. Every change is such that it allows the substance to fulfill its needs for the maintenance of its entity under new ambient conditions. This example might seem trivial as it describes facts well known from the field of physics. They are illustrated by routine observations of natural phenomena, such as the changing forms of water when it turns into ice, a liquid, or steam, depending on the temperature.

When the ambient conditions change, the *living* creatures alter their conditions of existence so that needs directed to the maintenance of their entity could be satisfied in a possibly comprehensive manner. The character of the transformation depends on the capacities and experience of individual organism. In some cases the change of the state is manifested by a change of the needs.

Dr. Alexander Yabrov

At lower temperatures, bacteria slow down their metabolism, so do the tress in winter—a period when they shed their leaves; or the hibernating animals, which restrict their activity. In its turn, this results in a change of needs. For example, the bear sleeps the whole winter in his den. His body temperature drops from 40°C to 4°C. This is a reflection of diminished rate of metabolism. Despite of an absence of supply, the hibernating organism survives because his needs for the maintenance of his entity are minimal. They are satisfied by the internal resources accumulated as a result of his adequate functioning during summer and autumn.

Some animals store a supply of food in addition to the increased inner resources. For example, in autumn a squirrel accumulates layers of subcutaneous fat, *and* stockpiles mushrooms and other food. The squirrel also makes the hollow cold-proof. At the onset of winter, the ambient conditions become harsh: it is cold and food is scarce. But this squirrel's own conditions of existence permit its needs for survival to be satisfied.

Migrating animals, such as birds, change their life conditions by relocating to places with a warm climate. Adaptation is a manifestation of the Fourth Law—animals adapt to their *needs*.

Man, whose capacities and experience are the broadest, uses all of the above methods to make his life conditions such that his needs are satisfied. Under conditions of deprivation, he slows down his physical activity. He also wears warm clothes in winter, and stockpile food, like the explorers of the North Pole, or the inhabitants of Greenland do. He may also migrate periodically like some Canadian retirees do, who live half a year in Toronto, Ontario, and half a year in Miami, Florida. In addition to this, man fine-tunes his conditions by using air conditioning, or by importing seasonal food from other countries.

The above examples are those of adaptation. The Fourth Law, thus, allows a more profound understanding of adaptation by bringing together the notions of *conditions* of existence (those ambient and own) and of the *needs* of maintenance of entity. We come to conclusion that adaptation is a form of adequate functioning (see Yabrov, 1987; 2001). It is not by chance—most of our examples illustrating the process of adequate functioning (above and also Yabrov, 1979; 2001; 2012) belong to the area of adaptation.

Yet, we are not quite sure that the above formulation of the Fourth Law conveys the idea of the conditions–needs interaction comprehensively enough. Further studies should show whether this is a satisfactory formulation. Existence is a new field. Reconsideration of the Laws might be needed. We view their role as the centers of condensation for further research in a new area. In our view, *needs is the leading factor.* An object adjusts to needs—first, rather than to conditions. Thus we change the currently leading principle of the theory of adaptation, which says that organisms adjust to conditions.

The following laws consider the role of *interaction* of objects from the point of view of their existence.

The Fifth Law: In Nature, Objects Exist Only In-groups

Natural objects functioning adequately exist in-groups to fulfill their needs in a comprehensive manner

Similar to the other laws, this law describes the existence of both animate and inanimate objects.

In Nature, we do not find a single physical particle, an atom of an element, or a single molecule of a substance—existing autonomously. Whether they exist in a solid, liquid or gaseous state, a group of particles, atoms or molecules is always involved. If we discover deposits of gold, or iron or any other minerals we do not find them as separate atoms or molecules, but as substances composed of many such atoms and molecules. Quarks represent an example of the elementary particles existing in-groups. A one exception: within the atom, electrons exist in isolation. By its uniqueness, exception rather confirms a law (see also Yabrov, 1012a).

Analysis shows that organisms exist in-groups to satisfy their needs comprehensively. Consider the following illustration from my experience as a virologist and cellular biologist. In virology, the most convenient experimental model is tissue culture. Living cells, e.g., human cells from foreskin, are grown out of the organism in a test-tube. Having a monolayer of living cells, an experimenter may infect it with the virus and study how the viral infection influences human cells without actually infecting any animals. Further, using the same method of tissue culture, it is possible to study how various protective substances, e.g., interferon, act against the viral infection.

In order for the cells to grow *in vitro,* i.e., out of the body, certain conditions should be maintained, such as constant temperature of 36°C, nutritional growth medium prepared by a special recipe that includes particular organic and inorganic components, etc. But, in spite of all these conditions, the culture will not grow if too small an amount of cells were used initially. Experimenters found that there is an optimal concentration of cells to start the culture. The lower the number of cells per milliliter of the medium, the more difficult it is to have them survive and multiply in a test-tube. This shows that for one's survival, growth, and multiplication, the individual cells need the others. The following experiment proves that this is the case.

If you add into the growth medium, say approximately 10% of liquid from an actively growing tissue culture—then this *conditioned* medium is able to maintain growth of a culture started with a low amount of cells. This experiment shows that grown together, cells produce some substances, which are necessary for their survival and multiplication. A single cell or a suspension with too low a number of cells is unable to provide for sufficient concentration of these vital substances.

We suggest that organisms exist in-groups by necessity, which can be explained as follows. In order to exist as an entity, an individual organism should fulfill certain needs. Some of these needs it *can* fulfill by itself and it does this according to the previous laws. But there are also such needs, which an object *cannot* fulfill alone; involvement of others is necessary for that. This refers to every individual member of the group. Therefore, in order for everyone to be able to maintain one's entity, each should also fulfill the needs for the others. Thus, natural objects exist in-groups because they *depend* on each other's aid in order that the needs directed to the maintenance of entity of each individual be satisfied comprehensively (Figure 8).

FIGURE 8. Interrelation of Objects in a Group (modified from Yabrov, 2001).

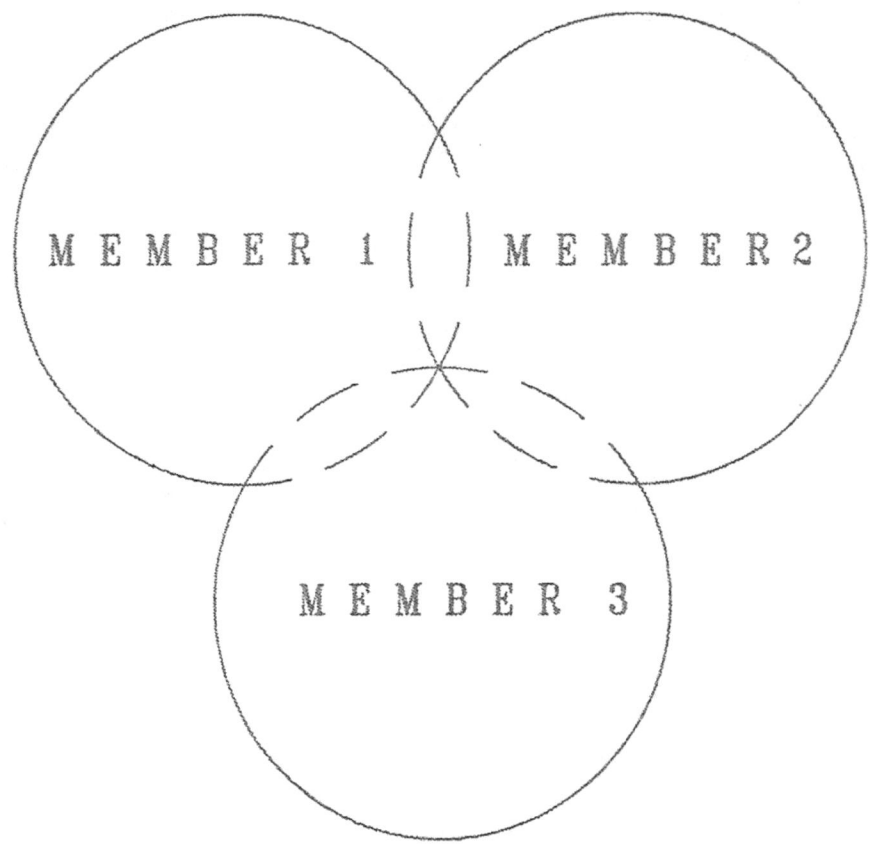

A group of three objects is presented—all are members of this group. Each one must satisfy certain needs to exist, some of which can be realized by the individual (continuous arc), some he cannot fulfill alone (broken arc). The latter needs are fulfilled together with the other members, while this individual satisfies needs of other members, which they cannot fulfill alone.

We can isolate a particle, an atom or a molecule of a substance and maintain it at experimental artificial conditions, but in natural conditions they do not exist alone. The same is valid for the living creatures—microorganisms, plants, and animals, including man.

Dr. Alexander Yabrov

It is possible to isolate a living being from members of the same species. Domestic animals are an example. Their existence, however, is *dependent* on man. Left alone, these animals cannot survive for any length of time if they do not join a group.

The aforementioned refers to Man: he lives in a society; alone he perishes. This statement applied to prehistoric man who could not survive alone, being too weak to hunt or to defend himself against various predators. In a contemporary setting too, without the society of others, modern man is as helpless as his prehistoric counterpart, may be even more so because the separation of labor in an advanced industrial society reinforces the mutual interdependence of individuals.

Extensive pertinent observations led me to a definitive conclusion that the fact of the natural objects existing in-groups is a Law of Nature. This fact prompted our studies concerning objectivity of description of reality based on the Copenhagen interpretation which deals with isolated particle (see further—Consequences and also Yabrov, 2012a).

The following law details the behavior of individual objects—primarily the organisms—in a group.

The Sixth Law: Law of Interaction of Objects In-groups

*In a group, an object functioning adequately exists by satisfying its needs and those of others on whom the maintenance of **his** entity depends, while these objects satisfy the needs of the former*

The 6-th law explains that in-groups organisms exist via interaction and cooperation.

According to the 1st Law, organisms exist by satisfying needs directed to the maintenance of entity. I intentionally did not state that the individual organisms satisfy *their* needs exclusively. Observations show that in groups organisms interact. They satisfy *their* needs, *and* those *of other* members of the group, while the latter meet those of the former. This is how each of them and the group itself exist.

This is emphasized in the Sixth Law, which asserts that besides *its* needs an object satisfies those of *others* on which the maintenance of its entity depends. This refers to each member of the group. This law describes functional interrelations among individual organisms forming a group. These interrelations are characterized by *interdependence* in preserving their entities. We call this law a *triad of interdependence* (Figure 9).

FIGURE 9. A Triad of Interdependence (Yabrov, 2001).

In **Figure 9**, the interrelations of adequately functioning individuals in a group are schematically described. A person located in the centre functions to satisfy his needs (the hand pointed toward himself). Simultaneously, he fulfils the needs of others (the hand pointed at others). While the others (who also satisfy their own needs) function to meet those of the individual at the centre. Figure 9 depicts the interaction of men, but the law is valid for any organism existing in a group.

In the multi-cellular organisms every cell consumes and processes structural materials, water and minerals. The cell exists by satisfying its needs, and at the same time, every cell, being a component of a tissue and of an organ, fulfils certain needs of the organism. In other words, it functions in accordance with needs of other cells of the same organism, which, in turn, fulfill those of this cell, as well as of the others. This is how the cells exist as parts of an organism, and how an organism exists as a whole. As the reader could see, this regularity plays the leading role in the maintenance of health.

In a group of animals, the same regularity is manifested by a separation of functions. A leader bears responsibility for the other members, for example, by keeping order, selecting where the group goes, where it stays, etc. Those with sharper senses watch while others sleep. These are illustrations of the triad that characterizes the regularity under discussion. An individual satisfies his needs and also those of the others, while the others fulfill this individual's certain need.

An especially sophisticated specialization of functions takes place in human society. The prehistoric tribe consisted of food gatherers, hunters, warriors and leaders. With the development of knowledge, professions demanding particular skills were established, such as those of a healer, or a tool-maker. This development is still going on today. For example, the profession of a computer programmer is a new one.

As a result of development of labor relations in a contemporary human society, the regularity that we now discuss is most obvious. An individual achieves satisfaction of *one's* needs by doing work for *others.* While the *others,* in their turn, do work for *this* individual. A doctor treats others, not oneself; a teacher teaches the children of others; a restaurant chef cooks for others; who, in their turn, make shoes, sew clothes, or do a multitude of other things useful for this cook, and the teacher, and the doctor, etc. The law emphasizes that it describes existence of the individual organisms that are acting *adequately.*

As stated in the Third Law (Article I), the phenomena of existence represent a spectrum of adequate and inadequate functioning, so that examples can be observed when the individual organisms in a group function *inadequately,* i.e., not according to the triad. In this case, the state of all the individuals worsens, and the longevity both of the individual and of the group shortens.

We may consider the example of the interrelations of cells in a multi-cellular organism. The cells function according to the triad.

Every cell satisfies its needs. It consumes and processes water, oxygen, and nutrients. At the same time, every cell functions as a part of the body. Liver cells process food, which is then distributed through blood to the other cells. Leukocytes produce antibodies, which protect these and other cells from infectious agents. Lung cells process gases—a process necessary for all the body cells. Thus, each cell satisfies its needs and those of the others. While the other cells function to satisfy those of other ones.

If one considers cancer cells, however, they satisfy only their needs, and multiply at the expense of the other cells. Eventually, the state of most of the cells worsens and the organism dies, and the cancer cells perish with it.

Everyone is familiar with examples of the deviation from the triad of interdependence at the social level of man's existence. It should be emphasized that the satisfaction of one's own needs—in-itself—is not a manifestation of an inadequate functioning, for an individual cannot exist if his needs are not being satisfied. Functioning becomes inadequate when an individual disregards his interests or those of others, or both. The Fifth and the Sixth Laws speak of the necessity and regularity of *interaction* of organisms. Only in-groups the needs of individual organisms could be satisfied comprehensively—the 5th Law. This is achieved via *cooperation* among the members of a group—the 6th Law.

The Laws Reveal a *Tendency*

One might argue: "The laws of Existence could be applicable for the living organisms, but do they describe existence of the inanimate objects as well?"

We should start with the acknowledgement that these are *new* laws. The laws of existence of the living organisms were *not* known. Thus we speak of the new laws in principle. Here they are discussed as the *general* laws of existence of natural objects—animate and inanimate. This inevitably provokes confusion. We are not used to think uniformly about *natural objects* not differentiating whether they are living or not living. Yet we want to understand how all objects—the animate and the inanimate—exist.

Dr. Alexander Yabrov

As a way to overcome the psychological obstacle, we would suggest an example of how we judge about behavior of various animals: We use our knowledge of the behavior of humans in a comparable situation. And vice versa: We apply knowledge obtained in the experiments with animals and extrapolate it upon humans. The goal of these comparisons: To understand *how*.

In case of the general laws of Existence our objective is the same—to understand *how*. Similarly to the above example, the method of comparison is exploited. We discovered the laws of Existence of organisms and applied this new knowledge for understanding of existence of inanimate objects (Yabrov, 2001). There is no absolute analogy. Yet these *general* laws of existence *are* applicable—they help us to understand *how*: they reveal a *tendency* common for all objects— that of the *adequate functioning*.

The following considerations by Lewontin (a geneticist) are apt here.

"One consequence of the value placed on great generality is that there is necessarily a tenuous connection between what has actually been observed in the world of physical phenomena and the theoretical claim. Between the idea and the reality falls the shadow of abstraction"—says Lewontin (2003).

He gives the following example:
"Newton's First Law, that bodies at rest tend to stay at rest and bodies in motion tend to stay in motion in a straight line unless perturbed by an external force, could not possibly have been a generalization of the motions actually observed by him. Neither he nor any other seventeenth-century observer ever watched a body move in a perfect vacuum with no external forces operating on it" (2003).

Yet, the contemporaries have gained from the laws of mechanics through understanding of the tendency, or the character of motion of physical bodies.

"The secret is in the word 'tend"- says Lewontin—"Tendencies are not observable. They are an abstraction around which the observations of the actual movement of bodies in different circumstances can be organized, in an attempt to understand the 'perturbing' forces" (Lewontin, 2003).

The laws of Existence reveal a common tendency toward adequate functioning.

How Nature Exists as a Whole

Nature as an object in-itself

How do the things and Nature exist? This is the problem initiated by metaphysics. The answer to the first half of the question—how the individual objects (things) exist—is given by the theory and the laws of Existence presented above. Now we approach the question: How does *Nature* exist as a whole?

To tackle the problem, we use our understanding of Existence of the individual objects. Nature as a whole—itself can be considered as a natural object, whose existence is maintained by certain processes and mechanisms. How it occurs?

When we discussed existence of individual objects, our emphasis was upon interaction and mutual influence taking place among the structural-functional levels *within* an object. In case of Nature as a whole, we rather concentrate our attention on the interaction and mutual influence *among* the objects. Our solar system abounds in the manifestations of the interactions and mutual influences of natural objects. Planets interact with the sun—this way the solar system exists. Moon influences the ocean tides. Meteorological events are a consequence of mutual interactions and influences of various objects and of the results of their functioning, such as the water and air currents having different temperatures, participation of solar radiation, and influence of vegetation upon humidity of air, formation of clouds, etc. A well known allegorical example is appropriate here of an influence of a flap of a butterfly's wings upon the weather. Another good example is presented by the discussed above Gaia theory, which asserts that organisms forming biocenosis themselves maintain the livable conditions on Earth as a result of activities and mutual influences of the organisms and the inanimate agents forming the biosphere (Lovelock, Margulis, 1974). Lovelock—author of the Gaia

Dr. Alexander Yabrov

theory—predicted that no life should be found on Mars because it was impossible to detect an uneven distribution of gases in the Martian atmosphere, which should have been caused by the activity of unevenly distributed interacting living creatures (Lovelock, 1988).

In order to give a scientific answer to a *how*, it is necessary to discover the processes and mechanisms responsible for the phenomena under study. Analysis of various examples of existence of Nature led me to conclusion that all the *fundamental* processes—those of *adequate functioning, physical-chemical interaction* and *evolution*, and their corresponding *basic* and *particular* processes and mechanisms—are involved in various manifestations of interactions and the mutual influences of different natural objects—this is how Nature exists as a whole. Since we speak of the *maintenance of entity*, we qualify the action and interaction of different fundamental processes and their mechanisms in the overall process of maintenance of existence of Nature as a manifestation of *adequate functioning*.

General analysis allows suggesting that in cases of existence of the objects, which have a limited time of existence, such as, for example individual organisms, existence is provided solely by the process of adequate functioning. The process of evolution needs a longer time to manifest itself than the limited time of existence of a single organism; whereas the duration of Motion in its terrestrial manifestations is brief, if compared with the time of existence of a terrestrial object. Example—diver falling into the pool. The period of his fall is so brief that we may disregard the concomitant manifestations of the process of existence, e.g. that the diver became older during time of his fall into the pool. And, of course, the duration of the fall and of existence of the diver is incomparable with that of evolution of Homo sapience.

When we deal with the manifestations of existence of Nature as a whole, which do not have time limitation, we consider also role of the other fundamental processes in addition to that of adequate functioning. Thus the process of evolution underlies development and existence of the species of organisms, and also the changes of celestial bodies. Whereas the process of physical interaction with its mechanisms is responsible for existence of the inanimate objects—terrestrial and celestial.

We have come to conclusion that Nature is represented by *all* the phenomena described by the fundamental notions of Existence, Motion and Origin. Therefore we cannot limit our explanation of how Nature exists by consideration only of the process and mechanisms of *adequate functioning* as such. Processes and mechanisms responsible for Motion and Origin should also necessarily be considered. These are the processes and mechanisms of *physical-chemical interactions* and of *evolution* (in some instances—revolution), correspondingly. *All* these fundamental processes and their mechanisms are involved in the maintenance of Nature's *existence*. Thus, we are confronted with a paradox. Nature exists by functioning adequately. In the case of Nature, however, existence is provided not solely by the process of adequate functioning, but by the harmonious interactivity of this process in concert with those of physical-chemical interactions and evolution (Yabrov, 2001).

This was Bohr who said: "How wonderful that we have met with a paradox. Now we have some hope of making progress". We suggest that the problem of existence of Nature can be resolved by using the concepts of the States of Objects and Nature, and of the State of Existence as a leading state. First of these concepts says that objects and Nature, or Physical Reality, abide in the states of Existence, Motion and Evolution simultaneously. This means that the fundamental processes responsible for these states are acting simultaneously. The other concept says that Existence subsumes Motion and Origin as its mechanisms. This means that the fundamental processes of adequate functioning, physical-chemical interaction and evolution—while acting simultaneously—interact among themselves in a certain manner. This interaction of the fundamental processes is a mechanism and a manifestation of *adequate functioning* of Nature as a whole. Interaction of the fundamental processes in the maintenance of Existence of Nature we have named—The Principle of Adequate Function (Yabrov, 2001). It is said: "Physics is the interrelationship of everything". The principle of adequate function explains how the in-

Dr. Alexander Yabrov

terrelationship comes to be: via interaction of the fundamental processes by the principle of adequate function.

One of the most articulate opponents of metaphysics and of the unified approach to understanding of Nature—philosopher-positivist Alfred Ayer, asked:

"How would he [metaphysicist] set about depicting the whole reality except through the depiction of its parts?" (1960).

Our argument to this is as follows: if one takes the parts of a wristwatch, their depiction will not be a description of the watch as a whole. We will merely obtain an inventory of its parts. The point is that an object comprises *structure* and *function*, while the parts represent the components of structure. Without consideration of the whole, the function is lost.

We explain existence of Nature as a whole as a result of interaction and mutual influence of natural objects, which is provided by the concerted participation of all the fundamental processes and their mechanisms. Function works through structure. Harmonious participation of the fundamental processes—those of existence, interaction and evolution—is a manifestation of *adequate functioning* directed toward and resulting in the maintenance of Nature as an entity in a favorable state possibly longer. Indeed, we deal with a very effective adequate functioning: By *our* time-scale, Nature exists for a long time—this is a fact. We come to conclusion that existence of Nature as a whole can be understood and explained via consideration of the processes and mechanisms maintaining Nature as an entity (Yabrov, 2001). Thus we found out how individual natural objects *and* Nature exist—it all comes to the adequate functioning directed toward the maintenance of entity.

PART 10.
THE
CONSEQUENCES

Conversation with the reader

So far, three scientific views of Nature were known: mechanistic, evolutionary and quantum mechanical. Knowledge of each of these views had consequences. We understand more and use this comprehension in our practice and for our further studies. We have discovered Existence an independent fundamental state of Nature, and described a new scientific view of Nature—that of *Existence of objects and Nature*. We also discovered the *fundamental process* that governs Existence, mechanisms of this process and its regularities—the Laws of Existence. This new knowledge has its *consequences*. Examples of these are given below. It should be emphasized that it is impossible to *foresee* all the consequences of a new view of Nature immediately after its discovery. Copernicus and Newton did not foresee the industrial revolution, which was a consequence of the mechanistic view of Nature. Neither Plank nor Einstein could predict tremendous development in the areas of information and communication, or invention of the computer—consequences of discovery of the quantum mechanical view. We may predict with certainty that the new view of Nature should bring many new discoveries. Still some consequences are obvious today.

The fundamental view of Existence is all-embracing—it unifies the animate and inanimate phenomena—a consequence described above. Other examples of *application* of a new view are given below in a certain order: Life Sciences, Physics and the Answers to Millennial

Dr. Alexander Yabrov

Questions. Data are presented in a concise form—for more details see the following monographs (Yabrov, 2001; 2002; 2012; 2012a).

LIFE SCIENCES
Advancement of Biology

Some of the readers might be surprised to hear that currently Biology is in a state of a deep depression. The following words of an expert reflect the situation:

"Of all the sciences we have mastered as humankind, the one we are the least advanced in is life sciences. And I think it's going to be the core challenge of the 21st century" (Dr. Zerhouny—Director of the National Institute of Health, USA).

Discovery of the new view of Nature meets this challenge. The problem of Biology is that the evolutionary theory is being applied for the explanation of *all* life phenomena. A theory—any theory, however, can explain *only* the phenomena underlain by the process discovered by this theory (Yabrov, 2001). Darwin's theory discovered the process of Origin—evolution. It should be used for understanding of origin of species. This is precisely what constituted Darwin's claim (1859). But it is also used for explanation of the every day phenomena of *Existence* of the living creatures.

Think of it: We have evolved already as a species. The pertinent mutations have been selected and inherited. Our every day life is being guided by a *different* process and its mechanisms. This is the process of *existence*—that of adequate functioning. This could be realized only based upon a new view of Nature. We come to conclusion that Evolution of species is not the only fundamental principle of Life. *We have discovered another fundamental principle of Biology—Existence of individual organisms.* The attempt to explain our everyday behavior by relying upon the theory of evolution only distorts comprehension.

The following words by the Secretary of the Lyceum of NY Academy of Sciences are apt here: "Now I realize that existence of organisms is not a lesser miracle than evolution" (J. Kirman). The problem is analyzed in details in a separate monograph (Yabrov, 2012).

Development of Medicine

New principle of Biology—that of existence of organisms, allowed us to develop a new theory of pathology. Till today, Medicine relies upon a theory developed more than 150 years ago. Virchow—world-renown German pathologist—introduced an idea that disease starts from a cell (1858). This was a great achievement. Prior to this—for 2000 years—medicine relied upon Galen's theory, which suggested that disease starts in the body liquids—blood, lymph, bile and phlegm. Patients were treated with bloodletting and purgatives—to free them from spoiled liquids. In many cases such treatment severely worsened the state of a patient. This is why Leo Tolstoy wrote about his heroine that "she survived *in spite* of the efforts of her doctors". Virchow named his theory—*cellular pathology*. It says that disease starts from the *structural* cellular lesion.

Based upon new methods of study and experimentation, including tissue culture, electron microscopy and others, we have developed Virchow's theory further. I found that besides the structural, a disease has also a *functional* component. Therefore we named a new theory—*cellular-function pathology* (Yabrov, 2001a; b). It allows explaining development of a chronic disease (Yabrov, 2012; 2013; Yabrov, Okunev, 2004).

What is health?

The reader knows that an organism exists by functioning adequately to needs. Human cell is an organism—in itself. In order to exist, it should fulfill its own needs and the needs of the organism. In other words, adequate function of a human cell is dual (Fig. 10).

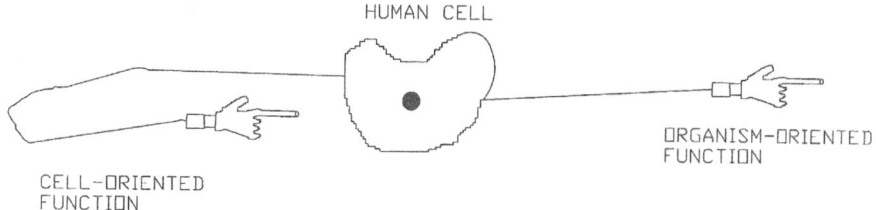

HUMAN CELL

ORGANISM-ORIENTED FUNCTION

CELL-ORIENTED FUNCTION

FIGURE 10. Dual function of a human cell (Yabrov, 2001a)

Dr. Alexander Yabrov

As shows Figure 10, the function of a human cell is dual—it is cell-oriented and organisms-oriented. Proceeding from understanding of necessity of fulfillment of both kinds of cell function, we come to the following definition: *Health is a state of an optimal balance of satisfaction of the needs of the cells and of the organism* (Yabrov, 1987).

How the chronic diseases develop?

During the passed half of a century, chronic diseases became the prevailing medical problem in industrially advanced countries (Yabrov, 1985). Modern medicine is not equipped to deal with this problem effectively. A general mechanism of chronic diseases is not known. Nobody even thinks about such a mechanism. Doctors are dealing with different isolated diseases. Because of narrow specialization of physicians, the patient deals with a different doctor each time when the problems arise. Doctor and patient are increasingly drifting apart precisely at the time when a continuous individualized medical observation becomes a necessary major feature of health care in our long-living society (Yabrov, 1985; 1986b).

First what is needed in this situation is a new theory of physiology and pathology that should allow a physician understanding the trend of the processes going on in the body continuously throughout the entire life of an individual.

We have found that *chronic* disease develops primarily as a result of disturbance of *function* of the cell: as an imbalance in the satisfaction of the needs of the cell and the needs of the organism occurs. When a cell fulfils primarily the needs of the organism—it occurs at the expense of one's own needs. It happens because the pool of resources and machinery in the cell is common both for its cell-oriented and organism-oriented functions. The cell suffers—its cell-orient function becomes increasingly compromised. As a matter of fact, the disease has *started* at the cellular level. But a person does not complain because the needs of the organism are satisfied, therefore the symptoms at the organ-organism level are absent. The complaints and symptoms will appear much later—when cells, whose needs have not been sufficiently satisfied for a long time, deteriorate and eventually become unable to fulfill the needs of the organism (Yabrov, 1987a).

Cellular-functional theory allows differentiation of the initial *functional* stage of a chronic disease. This truly early stage does not manifest itself by overt symptoms at the organ-organism level. In other words, the structural-functional theory explains that a person who has no complaints and symptoms (the one, whom today we consider to be healthy) might be ill and needing treatment. It is clear that the treatment applied at the functional stage is likely to be more effective. It may *stop and even reverse* development of a chronic disease.

A new—really early—stage of a chronic disease

We thus discovered a new stage of a disease—impairment of the cell-oriented function of a cell. This in itself is a serious finding. Medicine as a profession has existed for more than 2000 years, and as a branch of modern science—for more than 100 years, during which all the stages of a disease have apparently been discovered. The point is that, in order to catch an actual early stage of a chronic disease, we need to monitor health of cells as living units in-themselves and to look for the disturbance of their *cell-oriented* function. This discovery changes our approach to prophylaxis, diagnosis and treatment (see separate monograph—Yabrov, 2012) (N).

A NEW VIEW OF PHYSICS

Conversation with the reader

New view of Nature is not limited by the life phenomena. Inanimate objects *exist*. We have found that the fundamental process responsible for their existence is the same as for the living organisms. It is the process of adequate functioning; it unifies the animate and inanimate worlds in a united existing World (Yabrov, 1986). We also found that the Laws of Existence are *common* for both the animate and inanimate objects (Yabrov, 2001). Therefore the new view of Nature broadens and reforms our views of physical phenomena. We consider them as the manifestations of both Existence and Motion— therefore *both* physical laws and the laws of Existence are valid for understanding of physical phenomena (see Yabrov, 2012).

Dr. Alexander Yabrov

Matter, Antimatter and Existence of Physical Reality

It has been discovered recently, that in spite on the seemingly unshakable symmetry principle, a slight *excess* of matter over antimatter—within a one part in a billion—exists in Nature (Fitch, 1981; Cronin, 1981). It allowed Val Fitch—co-discoverer of the initial imbalance—to say that this slight asymmetry of matter and antimatter in the post-Big-Bang soup is "the reason we are here". Utterance by Marcus Tullius Cicero—the ancient Roman thinker—is appropriate here: "Nature abhors annihilation". For proving this scientifically—Fitch and Cronin were deservedly awarded a Nobel Prize.

This example is an indisputable illustration of the prevalence of the *tendency* toward Existence in Nature—Existence of the universe and of Man are the facts of reality (see Figure 1).

Three basic physical categories which assure Existence of the World

Proceeding from the facts, we come to a figurative generalization that *Existence* is the "point", or the "purpose of it all". The *asymmetry in favor of matter* v. *antimatter*—discovered by Fitch and Cronin, we consider to be the first physical factor—of three, which assure Existence of Nature in its complexity as it presents itself to us.

Another factor is the tendency to *self-organization*. For a long while, the leading scientific view of the direction of natural phenomena was—of everything tending to self-destruction and chaos. It was based on the concept of increasing entropy resulted from observation of behavior of particles in a closed system (Boltzmann, 1896). These observations have served as a basis of the theory of thermodynamics. However, in a general consideration of the trend of natural phenomena, which belong primarily to the open systems, this view has given way to the idea of self-organization.

Its initiation can be traced to the ancient atomists, and then to Descartes, who suggested tat the laws of nature tend to produce organization. Kant has applied the term "self-organizing", which later was used increasingly by the physicists, among them Norbert Wiener (cybernetics, 1961). The idea of self-organization acquired its broad application in the field of complex systems in the 1970s and 1980s. In 1977, Prigogine received Nobel Prize for his studies on the concept of self-organization within the framework of an idea of deterministic chaos. Currently, a serious contribution for substantiation of the role of self-organization has been made by Kauffman (1995-2008), whose broad experimental and theoretical studies provided a solid background for our inclusion of self-organization among the basic factors responsible for existence of our World.

Amazing is the interconnectedness of the ideas presented in this book with those—introduced by different thinkers in different times—all trying to describe, understand and explain the existing World. This is how Kant describes his vision of a self-organizing system in his *Critique of Judgment*:

"In such a natural product as this every part is thought as *owing* its presence to the agency of all the remaining parts, and also as existing *for the sake of the others* and of the whole, that is as an instrument, or organ...The part must be an organ *producing* the other parts—each, consequently, reciprocally producing the others...Only under these conditions and upon these terms can such a product be an *organized* and *self-organized* being, and, as such, be called a *physical end*." (Kant, 1790)

The reader may perceive in this description a close similarity with our explanation of existence of objects forming a group in the process of adequate functioning (see the 5th and 6th Laws). In light of this closeness of ideas, one might ask—what new brought the concept of a State of Existence and the corresponding theory? This is not just a matter of priority. The question is whether an *advancement* of knowledge and understanding of the basics of organization of Nature took place.

First of all, the theory of Existence has developed and changed the views of the natural phenomena. A new view considers the mechanistic-evolutionary worldview -initiated by Descartes, and by Lamarck—Darwin—to be of a limited embrace, which leads to incongruities. We have deciphered the mystery of Kant's thing-in-itself (see further). Our Laws of Existence provided the reasoning against the dominant view of the inevitability of the self-demolishing power of entropy. And we refuted Prigogine's view of Existence being a superfluous notion. In our dispute of 1998, Prigogine insisted that the notion of Existence is not needed—since all is described by the evolution of moving particles. I characterized Prigogine's description as "evolution-in—motion" (see above and Yabrov, 2002). It might be applicable for the description of some cosmic phenomena, or those which took place at the time of big-bang. But Prigogine's view cannot explain the ongoing phenomena of Existence of our world because different processes and mechanisms are responsible for evolution, motion and existence (Yabrov, 2001; 2002). Furthermore, as it follows from our theory—motion and evolution are the mechanisms of Existence.

Thus we come to the third pillar of existence of the World. None of the modern scientists considered and studied Existence as such. It was believed being beyond the scientific subjects. First who contemplated of Existence as a fundamental category of Nature—deserving special analysis independent from Motion and Evolution—were Einstein and Wheeler, I suggest. Our experimental results and theoretical generalizations in relation to existence of the living organisms were developed by me during 1969—2004. They were extended further upon the phenomena of existence of inanimate objects (1986—2001), and developed into a new view of Nature. In this volume introduced is the scientific concept of a State of Existence. We have developed a theory, which has explained how it takes place. Thus answered is the question "How come existence" posed by Wheeler (see further).

From all the scientific results known so far, we pin-point the *matter-antimatter asymmetry, self-organization and existence*—as the basic factors responsible for Existence of the World in its complexity as it presents itself according to our current knowledge.

David Gross says in his "Unified Theories of Everything":

"In physics there are two stages of understanding: You first ask the question: "How? how does it work?". After you have understood how it works, we are beginning to ask why?..." (Gross, p. 8, 1989)

We have answered How come existence. My interpretation of the meaning of Why—what is the purpose. Our answer: the point of it all is—Existence. It answers Leibniz' muse—"Why there is something rather than nothing?"

The universe is not pointless

Steven Weinberg—Nobel Laureate—one of the leading researchers in the area of quantum mechanics—said: "The more the universe seems comprehensible, the more it seems pointless" (1994). But the known *scientific facts* contradict this opinion, I suggest. Consider the living organisms—in most cases they function adequately in accordance with their needs so that they maintain Existence. Practice shows that life flourishes in spite of the adversities. These are facts of existence of *individual* organisms. Now consider development of species. Facts lead us to conclusion that evolution results in development of such organisms, which, on average, exist longer. We may conclude that evolution is directed, or has a tendency toward promotion of Existence. When conditions permit, the *inanimate* objects might exist for an indefinitely long time. As we mentioned, stone formations were found in Australia, whose geological age is close to that of Earth.

The Universe might be considered as pointless when the phenomena are viewed solely from the worldview of Motion (especially—motion of particles—the subject of studies by Weinberg). But from the worldview of Existence (Yabrov, 1979; 2001), we come to a different conclusion. It follows from the concept of a State of Existence that *the sense of it all is Existence.* Continuous motion and evolution maintain and promote *existence* of things and Nature. This is achieved via the process and the principle of *adequate function.*

The concept of a State of Existence subsumes that of an Anthropic Principle.

The idea of Existence being the "sense of it all" connotes the idea of the anthropic principle (Barrow, Tipler, 1989; Benioff, 2002). But an essential difference exists. The anthropic principle says: "The basic properties of the universe must be such that [intelligent] life can develop" (Barrow, Tipler, 1989); "Observers are necessary to bring universe into being" (Wheeler, 1996).

Dr. Alexander Yabrov

Benioff says:

"The basic properties of the physical universe must be such that a coherent theory is creatable. Since intelligent beings are necessary to create such a theory, it follows that the basic properties of the physical universe must be such as to make it possible for intelligent beings to exist".

And he adds:

"None of this implies that intelligent beings must exist, only that it must be possible for them to exist. Of course existence of intelligent beings is a necessary condition for actual creation of a coherent theory" (Benioff, 2002).

Benioff speaks here of a coherent theory of physics and mathematics "as an approach to a Theory of Everything".

Our analysis of the relevant works led us to conclusion that the researchers study different important problems whose resolution is helped by the acknowledgement of the anthropic principle. We, however, do not consider the presence of intelligent creatures, or any life forms to be a necessary condition for the state of Existence. To follow logic of the anthropic principle to the end, the Earth should have been the only planet in existence—the rest is just a supporting environment. If the anthropic principle where guiding the development of matter, the religious philosophy and the Ptolemaic astronomy, as well as the Copenhagen interpretation of reality, which place the Earth, Man and Conscience at the center of the universe, should have been reflecting reality. However, these concepts, though great, are merely the reflections of the current *knowledge* about reality, but not of the actual reality (see above). The principle of adequate function is broader than the anthropic principle. The concept of a State of Existence as initial and the eventual purpose and essence of all things embraces the habituated and the non-habituated Universe—it subsumes the anthropic principle.

Theory of Existence—the theory of Complex Adaptive Systems (CAS)

As usual, physics is at the forefront of development of science. A new science of Complex Adaptive Systems (CAS) aims at the study and explanation of both inanimate and animate phenomena. The new science is at a stage of collecting data by way of broad observations and the experiments in different areas. This is a usual way of development of a new science—it starts from accumulation of facts. However, for the establishment of a new science as a certain field of organized interrelated studies, a *general theory* is necessary. A theory designates the area of pertinent phenomena and determines the way of further investigations. This is achieved by the discovery of a general *process* underlying all the phenomena under study. As a matter of fact, this is the common underlying process which makes the phenomena *pertinent*. So far, the practice of a new science outstrips the theory.

Researchers working in the area of complex adaptive systems keenly realize a very limited applicability of the evolutionary theory in their field of studies. They speak of

"as-yet-undiscovered organizing principles in biology...except for Darwinian evolution." And they ask: "What if this view [of exclusivity of the evolutionary principle] is just a consequence of our inability to see?"

The same attitude is being expressed toward the physical laws:

"The dogma central to much of science that knowledge of the underlying physical laws alone is sufficient for us to understand all things is false." (Laughlin, Pines, at al, 2000)

The theory of Existence provides the answers to these questions. It explains that the Evolution of species is not the only fundamental principle of Biology. Another one is the Existence of organisms. It also discovers new laws of Nature side by side with the physical laws (above, and also Yabrov, 1979; 1986; 2001; 2002).

A common process underlying all the phenomena of CAS has not been discovered. Researchers in this field are guided by a certain functional feature common for the objects of their studies—objects of various complexities behave adaptively. This understanding, however, is insufficient to develop a basic theory. For comparison, consider a situation when in the absence of a theory of mechanics we would designate the objects of our investigations by the fact that they move. Or in the absence of the evolutionary theory, we would limit our observations of living organisms by an acknowledgement of their diversity. Now consider how far we have advanced as a result of discovery of the *fundamental processes* of interaction, evolution, and now—adequate functioning, correspondingly. We should repeat that recognition of adaptation is not enough for a general theory of CAS.

If we agree to consider the objects and phenomena belonging to the area of CAS from the point of view of their Existence (and not only Motion and Origin), we need a theory explaining how existence occurs. The Theory of Existence describes the process of existence, its mechanisms, and the laws. Here we concentrate attention on the *process of existence*. According to our theory, this is the process of adequate functioning—objects exist by functioning adequately to the needs directed toward maintenance of their entity (Yabrov, 1979; 1986; 2001). Consider again the examples of adequate functioning— above. All of them describe various cases of adaptation of organisms. Adaptation is a one of the *mechanisms* of adequate functioning.

We found that the process of adequate functioning is responsible for existence of both the animate and the inanimate objects. Examples of Gaia and of our solar system illustrate existence of the large-scale complex objects via adequate functioning (Yabrov, 1986). We suggest that other phenomena of existence of complex adaptive systems may get their explanation being considered as the manifestations of the process of adequate functioning.

Adaptation, thus, is a classic example of adequate functioning. It should be emphasized that the notion of "adaptation" itself needs essential reconsideration based on the theory of existence. Adaptation is being studied as adjustment to changing ambient *conditions*. It follows, however, from the 4th Law that the objects adjust to the *needs*—primarily. Conditions play their pole by influencing the needs.

Currently, the studies of CAS have acquired momentum of their own. An increasing amount of researchers of different specialties— physicists and biologists -enter this new important field of investigations. In order to unite their inevitably scattered creative efforts, a new *general* approach based upon the new view of Nature is necessary. The worldview of Existence gives physicists and biologists a *common* philosophical and scientific framework—a general approach to the study of complex natural phenomena. This unified view of Nature should serve as a base for a search of the new principles of the complexity science.

PART II.
SCIENTIFIC
ANSWERS TO
THE MILLENNIAL
QUESTIONS

Resolution of the problem of Existence

Conversation with the reader

The problem of Existence always occupied the leading minds of humanity. Yet, it remained unsolved. Now, having developed the concept, theory and the laws, we can explain *what* Existence is and *how* it occurs.

Spinoza—conatus

The first who noticed the phenomenon that objects preserve their existence was Spinoza (1632-1677). He suggested that all things, inanimate and animate, strive to maintain their being—a phenomenon, which was not obvious to anyone. This persistent and continuous strive—Spinoza named—*conatus*:

"The endeavour, whereby a thing continues to persist in its being, involves no finite time, but an indefinite time." (Spinoza, p. 15, E-III, 1951)

Dr. Alexander Yabrov

Being a lens-grinder, Spinoza came to his idea of conatus advancing from inanimate to animate objects, we believe. Being a physician—Yabrov came to a general idea of Existence and how it occurs—from observing the living organisms—first, and *then* the inanimate objects (this volume, and also Yabrov, 1979; 1986; 2001; 2012a).

Then the following objection might be suggested against the theory of adequate functioning. This argument is similar to what the opponents of the theory of gravitation insisted upon: "Nothing new was discovered [by Newton], since it was known from the time of Aristotle that things tend to the center of the Earth" (see Feynman, 1990). Analogously, some might ask: "What new did the theory of adequate functioning discover in addition to Spinoza's conatus?"

Spinoza observed the phenomenon, but he did not discover the process and mechanisms of conatus. He ascribed the persistence of being to the "divine nature". In other words, Spinoza did not *explain How* come existence?

Conatus is a great philosophical idea. It should be considered as a precursor of the theory of adequate functioning, we suggest (Yabrov, 2001). Still, Spinoza's keenness of observation, acumen and the might of generalization should be highly praised.

What is Existence?

All the sciences—in the areas of both physics and biology—using their particular methods, study the same subject: How objects and Nature exist (above, and Yabrov, 2001). To tackle this central problem efficiently, we need a definition of the notion of Existence. Throughout this book, we use the notions of Being and Existence without deliberating their difference; meanwhile these are *different notions.*

By our definition:

Being is a unity of structure and function.

It is a static *description* of physical reality, or Nature (see also further).

Existence is a dynamic *explanation* of how Being occurs.

How we *define* Existence? Existence is a complex *scientifico*-philosophical notion—more complex than that of Being, therefore we give it more than a one definition—all related to its *scientific* aspects.

1. *Existence* is a leading physical (material) *State* of the natural objects and of Nature.

This definition elaborates *scientifically* the Parmenides' concept "all is one". The scientific aspect is presented by our discovery of the *process* and the *mechanisms* responsible for the State of Existence. This definition uncovers to us the entire world—the *existing* World—inanimate and animate—in its unity.

2. *Existence is a unity of structure and function in the process of adequate functioning.*

This definition explains *how* Being occurs, or how things exist—via the process of adequate functioning. In other words, Existence explains the *dynamics* of Being. The definition reflects that Existence is a more complex concept than that of Being—the concept of Existence thus *subsumes* that of Being.

3. *Existence is a capacity of an object to maintain its individuality for an appreciable period of time.*

This definition is a further development of the previous one. It describes how the state of Existence manifests itself. To maintain one's individuality—entity, this is what it means to be in a *state* of existence—to exist.

Note: The above definitions reflect the essence of the idea. But the subject is extremely complex and multifarious. Probably, other definitions are also possible.

We have defined what Existence is. Now we should answer how Existence occurs, or using Wheeler's words: "How come existence?"

We have answered the question: "How come existence?"

Wheeler dedicated the book *Geons, Black Holes & Quantum Foam—A Life in Physics*—"To the still unknown person(s) who will further illuminate the magic of this strange and beautiful world of ours by discovering How come existence?" (Wheeler, Ford, 1998). Wheeler has suggested that this problem should be solved in 10, 100,

Dr. Alexander Yabrov

or 1000 years (1996). This prognosis implies that the problem of Existence is so important that it is worthwhile studying for 1000 years.

In his *At Home in the Universe*, Wheeler says:

"A single question animates this report: Can we ever expect to understand existence? Surely someday, we can believe, we will grasp the central idea of it all as so simple, so beautiful, so compelling that we will all say to each other, "Oh, how could it be otherwise!" (1996).

The problem is *solved*. The "central idea of it all" is that every individual natural object—as well as Nature as a whole—*actively maintains* its entity. And thus everything *exists*, which answers the question—"How come existence?" Existence is achieved by an active function adequate to needs directed toward the preservation of entity—*the adequate function* (functioning). *Certain* processes and their mechanisms provide for this function. In case of the individual objects, these are the process and mechanisms of adequate functioning. Nature as a whole maintains its entity through harmonic interaction of all the fundamental processes—those of adequate functioning, physical-chemical interactions and evolution, and their corresponding mechanisms. This harmonious interaction of the fundamental processes and their mechanisms, resulting in the maintenance of existence of things and Nature, we have named the *principle of adequate function* (Yabrov, 2001).

Our solution is simple and comprehensive—*things and Nature—themselves—actively maintain their entity and thus they exist.* Because of its simplicity and the brevity of its formulation, we might say that it is elegant, hence—beautiful. And it could *not* be otherwise. Objects and Nature exist—a fact. Every object has *its* entity. There is no other category to "know" the entity of an object and to have the capacity to maintain it—but this object itself. The same refers to Nature as a whole.

The following utterance is appropriate here:

"Through Einstein and his followers—Dirac, Murray Gell-Mann, Feynman, and others—it has become an article of faith in twenties-century physics that if a theory is simple enough, comprehensive enough, "beautiful" enough, it must be right" (Wheeler, Ford, p. 355, 1998).

Einstein ones asked whether God could create the World differently. If the aim was that the World and things *existed*—He probably could not make it differently. Or—to put it in a less categorical form—*He* has chosen the most simple and reliable way: Existence via adequate functioning. Spinoza states in *The Ethics:*
"Nothing in the universe is contingent, but all things are conditioned to exist and operate in a particular manner by the necessity of divine nature" (1951).

The theory of Existence has revealed this "manner"—this is adequate functioning.

Our solution is simple, but it is far from being obvious
It is appropriate to start this section with the words by Galileo Galilei: "All truths are easy to understand once they are discovered; the point is to discover them".
It needs to be emphasized that our explanation is simple, but it was not obvious prior to its discovery—see Wheeler, 1996; Wheeler, Ford, 1998. On the contrary, rather the *opposite* was considered to be obvious. Let's analyze the following example.
Hermann Weyl in his *Philosophy of Mathematics and Natural Science* (1949) has stated:
"The objective world simply *is*, it does not *happen*".
Frank Wilczek, who cited this statement, exclaimed: "Parmenides loves that, and so do I" (p. 278, 2006). However, in spite of the support of such authorities of antiquity and modernity, as Parmenides and Wilczek are, Weyl's claim is *wrong*. The objective world does not "simply" exist ("is"). Its existence is provided by a certain fundamental process and the mechanisms. And there are Laws of its existence. In other words—the world *happens*. How does it happen?—This was the essence of Wheeler's question "How come existence?" The answer needed development of a theory discovering the underlying process, mechanisms, and the laws. Says Einstein:

Dr. Alexander Yabrov

"When we say that we understand a group of natural phenomena, we mean we have found a constructive theory that embraces them" (Einstein, p260, 2000).

This is what our study has achieved.

We hear an argument by the experts in the theory of entropy: "Everything tends to self-destruction: It is enough to apply force and any object tumbles down". Our counterargument: "This is the point—you need to apply force. This means that the objects tend to *maintain* their Existence."

The scientific theory of existence discovered the key notions necessary for the scientific study of all the phenomena of Existence. In particular, it differentiated Existence as a notion independent of those of Motion and Origin. Then it defined Existence as a leading physical State of the objects and Nature. The theory discovered the fundamental process underlying Existence—adequate functioning. It also discovered the structural-functional organization of existing objects showing that existence is upheld by basic processes and mechanisms differing by their organizational levels. The theory discovered interrelations of the key notions of existence, and formulated them in the general laws of Existence. These new findings allowed us explaining scientifically—*how* things and Nature as a whole—Exist: we have answered "How come existence".

Spinoza's achievement was that he *observed* conatus. The phenomenon was not obvious at all. Nobody else saw it. Another example: According to our interpretation of Wittgenstein mystical allusions, this philosopher probably observed Existence, but he could not describe what he saw (see Yabrov, 2002). Einstein thought of Existence in a broad sense, which, in our opinion, happened to be the essence of discrepancy of his views with those of the majority. But Einstein used arguments and illustrations limited by the sphere of Motion. This prevented him from persuading his opponents. These examples demonstrate that the solution discovered by us—was not obvious. If now you see it as being obvious, this is an illustration of the simplicity and elegance of the solution that was found.

Pertinent to the above is the argument suggested by Professor Polsik of the Copenhagen Bohr Institute: "Objects maintain their existence by maintaining their existence—this does not sound per-suasive to me"—he raised an objection. In reply, I quote the Galileo's law of inertia:

"Everybody preserves in its state of rest, or of uniform motion in a right line, unless it is compelled to change that state by forces impressed thereon."

Says Einstein about this law:

"The discovery and use of scientific reasoning by Galileo was one of the most important achievements in the history of human thought, and marks the real beginning of physics...Once the idea of inertia is understood, one wonders what more can be said about it" (Einstein, Infeld, pp. 8 and 153 ; 1960).

We conclude: "Nothing more can be said: the state of inertia is the property of the body". The same refers to adequate functioning.

Phenomenon of the self-organization could be considered as the other example, besides that of inertia: Complex objects are being or-ganized by way of self-organization (see further and also Kauffman, 2008). That's how Nature works—very simple. The role of a scientist is to notice a phenomenon, explain how it occurs and then use the new knowledge. Now we broaden the latter example: The self-organized complex objects *exist* by functioning adequately. Thus by unifying phenomena of self-organization and adequate functioning we have explained—just now—how Order supersedes Chaos. Prigogine—a discoverer of "deterministic chaos"- spoke of spontaneous assembly of particles resulting in formation of structures, which then dissi-pate—to form the other ones through bifurcations. This is how he viewed Existence—as "evolution of particles in the process of mo-tion"; therefore he insisted that our interpretation of Existence was not needed (our dispute is described here and also in Yabrov, 2002). The point is, however, that to get to "life and art" (as Prigogine puts it)—the spontaneous structures should stop dissipating. They should keep their entity for an appreciable period of time. The Theory of Existence explains how Nature achieves this.

Hawking says in his *Brief History of Time*:

Dr. Alexander Yabrov

"A complete, consistent, unified theory is only the first step: our goal is a complete understanding of the events around us and of our own existence" (p. 69, 1988).

This is the task that has been fulfilled.

Wheeler is a scientist and a poet. It is appropriate to end the study of How come Existence by an *ode* to *Existence* written by Wheeler:

"Existence, the preposterous miracle of existence! To whom has the world of opening day never come as an unbelievable sight? And to whom have the stars overhead and the hand and voice nearby never appeared as unutterably wonderful, totally beyond understanding? I know no great thinker of any land or era who does not regard existence as the mystery of all mysteries" (Wheeler, p.184, 1996).

This mystery is resolved. Our view of Nature broadened and became clearer.

The new view of Nature allows us solving some problems which occupied the thinkers for hundreds and thousands of years.

Is Nature knowable?

Greek philosopher Plato (400-300 BC) approached the problem as follows.

In Plato's view—Nature is represented by the individual objects—things; a materialistic view. To understand Nature, it is necessary to know everything about an individual thing. As a result of his observations, Plato introduced a theory according to which, what we know about the things from our senses—what Plato called "appearance"—is *not* a comprehensive knowledge of reality. Things possess an attribute that we *cannot* perceive by senses. Without knowledge of this inner *essence* of things, which Plato called "Idea", our knowledge of reality is *incomplete*.

In modern time, Kant (1724-1804) corroborated the views of Plato concerning impossibility to know things comprehensively. In his *Critique of Pure Reason* (see edition 1997), Kant emphasized the difference between the appearance of a thing, and the "thing-in-itself", which is an *unknowable* residue of an object. Appearance, which is "empirically real," he called "phenomenon", while the component of an object, which is not amenable to our senses—thing-in-itself—he called "noumenon." Thus Kant's theory of the thing-in-itself is closely related to that of Idea by Plato.

The views of Plato and Kant about an *impossibility* of learning reality comprehensively had had a profound influence upon development of epistemology—the theory of science that investigates the origin, nature, methods, and *limits* of knowledge. Furthermore, it influenced essentially our entire worldview. In spite of success of science, *nescience*—the doctrine that nothing is truly knowable—holds an established position in the very psyche of the modern researchers and the lay public alike.

Guided by Kant's reasoning, many modern physicists consider *reality* being not knowable in principle. This refers especially to those studying quantum mechanics, which is illustrated by the following utterance by Bohr:

"There is no quantum world. There is only an abstract physical description. It is wrong to think that the task of physics is to find out how nature is. Physics concerns what we can say about nature" (see Heisenberg, 1971).

This view is reflected in the Copenhagen interpretation of reality.

We see that the problem of Idea, or Noumenon—i.e. the part of Nature that is beyond the grasp of our mind—has a very serious influence upon development of the philosophical *and* the scientific thought till today.

Resolution of the Central Problem of Metaphysics—Nature *is* knowable

We have classified the answering what essence lies in the objects besides the properties experienced through our senses—as the central problem of metaphysics (Yabrov, 2001). Our approach is based on the Theory and the Laws of Existence. The 2nd Law describes the *structural-functional* organization of existing objects. The Law says that objects comprise <u>structure</u> *and* <u>function</u>. Analysis based on this law led us to conclusion that the part amenable to our senses—Plato's "appearance", or Kant's "phenomenon"—is *structure*. Whereas the part of an object *not* perceived via our senses—the Idea, or Nomenon (the thing-in-itself)—is *function*. Though it is not sensed, the latter *is knowable* by way of the *scientific* methods of study. Consider the following examples.

We do not see the objects of the micro-world. But we may discover their existence, describe and explain their properties by registering the manifestations of their *functioning* using methods of science. As a matter of fact, all the discoveries in the area of atomic physics, among them discovery of an electron, of X-rays, and of radioactivity were achieved as a result of scientific investigation of the functioning of physical particles. A striking example of efficacy of this method of investigation represents discovery of the atomic nucleus by Rutherford. He could not see the nucleus itself. But he found that in the process of bombardment of mica or foil by the alfa-particles, on rare occasions (1 in 20.000 collisions), some particles bounced back. Rutherford described it "as if you fired a 15-inch shell at a piece of tissue paper and it came back and hit you". Based on the results of his experiments, Rutherford came to conclusion that there was a massive positively charged core inside the atom. This is how nucleus was discovered without seeing it. It is hard to imagine a more appropriate representation of the 'essence' of things—an inner component not amenable to our senses—than an atomic nucleus.

Examples of finding the essence of things and phenomena (not amenable to our senses) by studying the manifestations of functioning of the objects are not limited by the physical micro-world.

Astronomers use the signs of irregularities in motion of known planets in their search for the new ones—not visible directly.

Thus planet Neptune was discovered "on a tip of a pen"—by the method of mathematical analysis aimed at explanation of certain functional aspects—without observing the new planet itself. Description of the discovery presents fascinating reading, (We do not discuss here whether Neptune should be classified as a planet).

Consider now the living objects. We base our judgment about the state of health of a person primarily by the evaluation of functioning. Most of the diagnostic tests are the *functional* probes. It is not always necessary for a physician to see the damaged cells of the liver, or the inner surface of the vessels of a patient to diagnose hepatitis, or high blood pressure, or atherosclerosis. Doctor achieves this by monitoring the function. We may generalize that the entire area of diagnosis of diseases is based primarily upon studying of the functioning of the human body.

It worth noticing that in the examples analyzed above, the study of *function*—a part of a thing, which is not amenable to a direct perception—resulted also in discovery and a more profound knowledge about *structure*—the part that is supposed to be immediately perceived. For example, studying function, we come to conclusion that the walls of sclerotic arteries are rigid.

Thus, we have considered examples from the areas of physical particles, physical bodies, and the living organisms. Analysis of examples from different fields of the scientific study of Nature lead us to a one and the same conclusion—the part of a thing, which is inaccessible for a direct perception—is *function*. It is knowable via the methods of science. Thus we have eliminated the seemingly permanent rift between the knowable and the unknowable. We came to a definitive conclusion: Nature *is* knowable.

We summarize with the following statement that might seem to some to be too bold:

Our understanding of the *essence* of a thing, which is not directly perceptible through our senses, as a *function*—nocks out the base from under the seemingly solidly established view of an incapability of Man to comprehend Nature.

This is how it goes in the exploration of Nature—a problem remains unresolved for decades, centuries or millennia. Concepts and theories are being developed based on the view that the problem is unsolvable. And then—in a moment, everything changes (though the "moment" could be a result of long-term studies). Our horizon widens—a broad road for the new explorations becomes seen clearly. Examples of discoveries by Columbus and Mendeleyev come to mind. Now we shell discuss a different millennial problem—that of Being.

Dr. Alexander Yabrov

From the Central Problem of Metaphysics toward the Central Problem of Philosophy—that of Being

Being is a millennial problem. In our time, Heidegger characterized the question "What is Being?" as the central problem of philosophy (1927). Resolution of the central problem of metaphysics—that of Idea-Noumenon—creates a base for resolution of the millennial problem of Being.

Parmenides considered all natural phenomena as the manifestations of Being—"all is one", whereas Motion and Change were mere appearances. We have discussed just above the Plato's approach to the problem of Being. He studied things and tried to understand them. Eventually, he came to his concept that Being consist of two parts—those of appearance, which we can understand, and of an Idea—a non-understandable part.

Aristotle considered Being as an apogee of description of existing world—an essence that subsumed the elements. He posed the question: "What is Being?"—which he considered—will never be answered:

"Indeed the question that was raised in the beginning and is now and always being raised and never answered—namely, what is *being*?" (Aristotle, p. 25; 1943).

What is Being?—A 2000-year Conundrum is Resolved

We have solved the problem of the Idea-Noumenon—this is *function*. Now we come to the formulation of what is Being:

Being is a unity of structure and function (Yabrov, 2001).

By way of modern physics, this formulation explains the meaning of Aristotelian "essence" of Being. It also answers the inquiry of Plato and Kant about the composition of existing things. Instead of representing a unity of the knowable and of an *un*knowable, Being represents a *knowable* unity. We assert:

In Nature, structure does not exist without function, and function does not exist without structure.

Thus we have solved the central problem of philosophy. We have answered the question "What is Being?"- which remained unanswered for two millennia and was supposed never to be answered: Being is the unity of structure and function. This formulation leads us to understanding of the foundation of Reality

Structure-Function—the Foundation of physical Reality

The beginning of natural philosophy is being referred to the 6th -5th centuries BC. We calculate the age of modern science starting from the discovery by Copernicus of Motion as an independent fundamental notion (above, and Yabrov, 2001). Copernicus' *De Revolutionibus Orbitum Coelestium* had had seen the light in 1543. Thus, modern science nears its half-millennial mark. Based on the knowledge about Nature acquired by natural philosophy and modern science during two and a half millennia, we now approach the problem of a foundation of physical reality. But what should we consider as the *foundation* and where should we look for it in the overall framework of the human knowledge?

In his *The Quantum World*, Kenneth Ford says:

"The reason for quantum theory, if there is one, could come from below or from above. That is, it might be found in the subatomic domain of the smallest intervals of space and time, or it might be found in some cosmic principles governing the universe at large." (Ford, p. 221, 2004)

Ford speaks of the quantum mechanics. By suggesting two polar alternatives, Ford not only points to *where* the reason of quantum physics might be. He also implies *what* it could be: the smallest, short living material particle(s), or the greatest all-embracing idea(s)-concept(s) of physical reality. As far as it concerns the quantum mechanics, Ford's approach, which considers these alternatives, is justified. But when we are trying to formulate the foundation of Nature in general—the only alternative is "above": A broad-scale concept. Why could not it be a particle? Particles form the initial level of organization of all and every natural object. It might seem that this fact itself places particles at the *foundation* of Nature. We, however, consider this view to be too formalistic. We speak here not of "anatomical" foundation, but of a *conceptual* one. It follows from the principle of the structural-functional organization of existing objects (the 2nd Law) that natural objects, depending on their complexity, are composed of several levels of organization—from that of the physical particles to the social one. A foundation of Nature cannot be identified exclusively with any *one* of these levels. Moreover, it necessarily should include *all* the levels. This means that the foundation of Nature should be a *universal concept*, or as Ford puts it—a "universal principle".

Our approach to the problem of the Foundation of Nature

We are confronted with the dilemma of an *approach* to a problem under consideration. The long-term studies and contemplations led me to conclusion that in order to formulate a concept of the foundation of Nature, it is necessary to consider the meaning of the notions of Reality, Being and Existence. Within their realm lays the Foundation's disposition, we suggest.

A consecutive comparative analysis of these notions reveals a certain dynamics of their epistemic message. We perceive here a gradual transition from philosophy to science. *Reality*, though a material category, is essentially a descriptive abstract philosophical notion: it encircles the real things but does not tell us what constitutes them. It is *not enough* to realize that the things are real in order to understand their essence—i.e. matter, and explain how they exist.

The notion of *Being*, in our interpretation, explains the essence constituting things—we have discovered that this is the unity of the *structure* and the *function*. Thus we have discovered the essence of physical reality, or matter.

Some might argue that we speak of the known things. Notions of structure and function are well known. Though it is not particularly discussed, it does not provoke objections that they may be represented simultaneously in various phenomena. The point is, however, that we consider structure and function as the fundamental notions; and the *unity* of structure and function—as the essence of Matter.

The notion of *Existence* subsumes the former ones and explains *how* reality exists. Consideration of structure and function, and then of the processes and mechanisms of Existence—emphasizes the *scien-*

tific—explanatory (rather then descriptive)—character of the notions of Being and Existence.

The fact that we need both a philosophical and a scientific characterization of the realm where we are looking for the Foundation of Nature demonstrates that—at this level of abstraction—we could not (and should not) separate philosophy and science. Only via unification of these leading sources of objective knowledge could we discover the Foundation.

The Kernel of the Foundation of Nature

The study of a State of Existence and analysis of the philosophical inquiry into Being presented in this volume (and in the relevant materials described in our other publications) led me to an idea that these tightly related concepts—Being and Existence (philosophical and scientific ones)—constitute the kernel of a Foundation of Nature. Consider the following reasoning.

According to our definition, "Existence is the unity of structure and function in the process of adequate functioning" (above, and Yabrov, 2001). Following the same logical analysis, we come to conclusion that *Motion* is the unity of structure and function in the process of physical-chemical interaction; whereas *Origin* (Change) is the unity of structure and function in the process of evolution (in some instances—revolution).

The above definitions describe all the manifestations of the *states* of physical reality. And in all of them *structure and function* is the kernel—an *object* of action of all three fundamental processes and their mechanisms. Now we apply the principle of adequate function. It says that Nature exists via harmonious interaction of the fundamental processes, which is a manifestation of adequate functioning. Logical synthesis of the concepts of Being and Existence—as they are described and explained in this volume—form the *foundation of Physical Reality,* i.e. they lead us to the comprehension of *what* is Nature and *how* it exists.

Unity of structure and function is everything and every thing.

Everything and every thing exist via the process of adequate functioning.

Dr. Alexander Yabrov

It is clear now why the thinkers from antiquity to our days considered the question "What is Being" to be the center of the philosophical inquiry into Nature. And why Wheeler considers explaining "How come existence?" the central problem of modern science. Resolution of these problems leads a researcher to understanding of what constitutes the foundation of physical reality.

A great harmony of Being—Existence

Now, that we have explained what Existence is and how it occurs, we have an opportunity to review the panorama of human thought about the World, Earth and Man. We may perceive a great harmony of vision and understanding of Nature in its unity throughout the history of mankind. A mathematician and a physicist cannot and should not dismiss these considerations as merely "philosophy". We all constitute humanity. How the humanity sees the World—is our business, independently of the professional methods, which we exploit to study it.

We start with religion. It's central subject: Being. What the humans have seen always was—that the World—people, things and the universe—where there...they *were*. Yes, they were moving, and they were changing, but they—in the first place and most essentially—they existed. Existence of everything is the essence of Being. Humans always realized it—therefore Being was the central subject of religion. Then the era of natural philosophy has started. Parmenides—the father of metaphysics—has declared: "All is Being". We see the harmony of thought. Spinoza relieved the idea of Being from its divine component. He spoke of Nature, but he preserved the notion of Being via conatus. Now the modern science has undertook a study of this eternal problem. As described here and in our pertinent works: Being is a *physical state* of Existence of objects and Nature; a leading state, of which Motion and Evolution are the manifestations and mechanisms. Science explained *how* Existence takes place: It discovered the underlying *process, its mechanisms, and the Laws.* Analyzing these scientific discoveries, we come to realize that the fundamental vision of Nature, which humankind always had—that of Being—remains. A profound, all-embracing harmony and persistence of the human thought opens to our mind: A great harmony of Being-Existence. Wheeler's ode to Existence fits here again:

"Existence, the preposterous miracle of existence! To whom has the world of opening day never come as an unbelievable sight? And to whom have the stars overhead and the hand and voice nearby never appeared as unutterably wonderful, totally beyond understanding? I know no great thinker of any land or era who does not regard existence as the mystery of all mysteries" (Wheeler, p.184, 1996).

New discoveries should follow

The above consequences are only the beginning. Experience shows that a new view of Nature plays its cognitive creative role for ages. Who could foresee the impetuous development of industry at the time when Copernicus discovered Motion as an independent fundamental notion? Similarly, nobody foresaw the invention of computer at the time when Planck and Einstein published their first papers in the area of quantum mechanics. There are reasons to believe that the new view of Nature—the one of Existence of objects and nature—should bring new discoveries now and in the future.

PART 12.

WE ARE FACING A NEW FUNDAMENTAL VIEW OF NATURE

We are Missing Existence—a New Perspective

Thus, after 2500 years of continuous thought about the essence of Existence, humankind came to a simple, but profound understanding that Existence is a leading physical State of every creature, every thing and of Nature as a whole. What does this insight adds to our current knowledge? It opens a new fundamental view of Nature.

It becomes obvious that modern science *missed* the studying of a largest and central realm of Nature. This is a striking revelation: an entire—pivotal—area of natural phenomena still remains unexplored. It should be acknowledged that ultimately, all the sciences study Existence (Yabrov, 2001). Physics, chemistry, biology, and other sciences bring certain knowledge about Existence of objects— using their specific approaches and methods. But this knowledge is *indirect*—rather it is a particular fraction of our exploration. What has not been done—an immediate, direct study of the phenomenon of Existence as such aimed at discovery of its underlying process; its mechanisms; and its Laws.

Dr. Alexander Yabrov

In his book "The Troubles with Physics" (2006), Smolin speaks of the current problems of modern physics. He emphasizes that the pace of discoveries has slowed in spite of the intensive creative efforts. He concludes: "We are missing something Big". In other words, this researcher suggests that our efforts are misdirected. The same idea is reflected in the courageous words by Gross—Nobel Laureate in physics, 2004:

"We don't know what we are talking about...The state of physics today is like it was when we were mystified by radioactivity... They were missing something absolutely fundamental. We are missing perhaps something as profound as they were back then". (quoted from Wilczek, 2008)

Even a deeper recession characterizes the state of the Life Sciences (Yabrov, 2001; 2012a). Theory of evolution is *correct*. But as any scientific theory, it can explain only the phenomena underlain by the process discovered by this theory (Yabrov, 2001). Current attempts to explain *all* life phenomena by the theory of evolution—mislead the researchers, and prevent advancement of our studies of the everyday life phenomena.

Thus, objective analysis of the current state of science leads us to conclusion that the basic science—both Physics and Biology—is in a state of stagnation. To find the road to further advancement, it is necessary to think on a world scale. Discovery of a new fundamental view of Nature is in line with this order. Half-millennial history of modern science shows that the new fundamental *scientific* views of Nature are being discovered ones in 100 to 200 years. The current mechanistic-evolutionary scientific view of Nature does not embrace all the natural phenomena. *We are missing Existence.* A new fundamental view of Nature is described above. This is the View of Existence of Natural Objects and Nature. It paves the way to astounding advancement of science as did the previous fundamental views of Nature—mechanistic (Copernicus, Newton), evolutionary (Lamarck, Darwin), and the quantum mechanical (Planck, Einstein).

Triumph of science

Throughout the entire millennial history of human thought, the problem of Existence (Being) persistently was at the center of attention of Man. It was the major subject of religion and then of natural philosophy. No wonder: What else could be more important for Man, and for that matter, for other living creatures—but their Existence.

Now modern science took upon itself the task of solving the problem. Einstein and Wheeler initiated the scientific inquiry. Our studies allowed resolving the problem. The answer is given in this book. Consider again the dynamics of exploration (Figure 1). Natural philosophy starts by Parmenides' statement: "All is Being!" The father of metaphysics was right. Yet, an explanation was necessary. "What is Being?"—Aristotle asked. For millennia, the leading philosophers tried to find the answer—with no conclusive result. Science has approached the dilemma having already solved the problems of Motion and Evolution. These discoveries allowed mastering the scientific method. Based on these achievements, science initiated a search for an all-embracing explanation of natural phenomena. Einstein looked for the harmony in Nature. Wheeler pointed that it is in the notion of Existence. "In any field, find the strangest thing and then explore it"—Wheeler said (1957).

This book explains what is Being, what is Existence, and how they take place ("how they come?"). The solution is achieved based upon the methods of science guided by the worldview of Existence. It is found that Existence is a physical State of objects and Nature. Discovered are the underlying fundamental process of Existence— adequate functioning, its mechanisms, and the laws. A new view of Nature proves Parmenides to be right. The words of another ancient philosopher—Aristotle, who studied "Being as Being"—are appropriate for *this* study of *Existence as Existence*:

"Supreme among the sciences and superior to all subordinate sciences is that which knows the end for which everything takes place, which is the good for each thing and, as a whole, the highest good for all Nature" (Aristotle, p. 9, 1943).

Understanding of Being-Existence—triumph of science.

REFERENCES

Anastopoulos, C. Particle or Wave. Princeton University Press. Princeton, NJ. 2008.

Anderson, P. W. More is different. Science, pp. 393-396, v.177, No. 4047, 1972.

Aristotle. On Man in the Universe. Walter J. Black, Inc., Roslyn, NY, 1943.

Ayer, A. J. Language, Truth and Logic. V. Gollantz, Dover, 1946.

Ayer, A. J. The Central Questions of Philosophy. Weidenfeld and Nicolson. London, 1973.

Barrow JD., Davies PC., and Harper CL., Jr. Science and Ultimate Reality: Quantum Theory, Cosmology and Complexity. PCRS/Metanexus, 2004.

Barrow, J. D. and Tipler, F. J. The Anthropic Cosmologic Principle. Oxford University Press, Oxford, 1989.

Barrow, J. D. Theories of Everything. The Quest for Ultimate Explanation. Clarendon Press. Oxford, 1991.

Barrow, J. D., Davies, C. W., and Harper, C. L., Jr.—editors. Science and Ultimate Reality: Quantum Theory, Cosmology, and Complexity. Cambridge University Press, Cambridge UK, 2004.

Benoiff, P. Toward a coherent theory of physics and mathematics. Foundations of Physics, v. 32, No. 7, 989- 1029, July, 2002.

Dr. Alexander Yabrov

Bernard C. Lessons on the phenomena of Life common to animals and vegetables, pp 129-173. In: Langley, I, ed. Homeostasis. Origins of the Concept. Dowden, Hutchinson and Ross, Inc., 1973.

Bohr, N. The quantum postulate and recent development of atomic theory. Nature, pp. 580-590, v. 121, 1928.

Bohr, N. Can quantum-mechanical description of physical reality be considered complete? Physical Review, pp. 696-702, v. 48., 1935.

Boltzmann, Ludwig. *Lectures on Gas Theory.* Reprint, Dover, 1896.
Brill, D. Ph.D. Thesis, Princeton University, Princeton, NJ, 1959.

Carlson 1D. Physiology of exposure to cold. Physiology for Physicians, 1-7, 1964.

Carnap R. Quoted from Park, D. The How and the Why, p. 234, Princeton University Press, Princeton, NJ, 1988.

Cercignani, Carlo (1998). *Ludwig Boltzmann—the Man Who Trusted Atoms.* Oxford University Press. ISBN 0-19-850154-4.Cishing J. Quantum Mechanics: History and Copenhagen Hegemony. Univ. of Chicago Press. Clarendon Press. Oxford, 1991.

Clark RW. Einstein. The Life and Times, p. 609. The World Publishing Company, NY, 1965.

Cronin, James W. CP-symmetry violation: the search for its origin. Science, pp. 1221-1229, v. 212, No. 4500, 1981.

Curd, P.K. The Legacy of Parmenides. Princeton University Press, Princeton, NJ, 1998.
Darwin Ch. The Origin of Species by Means of Natural Selection. The Modern Library, NY, 1859.

Darwin Ch. Autobiography, 1876. Azreads, NY, 2009.

Davies, C. W. John Archibald Wheeler and the clash of ideas. In: Science and Ultimate Reality: Quantum Theory, Cosmology, and Complexity (Barrow, J. D., Davies, C. W., and Harper, C. L., Jr.—editors), pp. 3-26. Cambridge University Press, Cambridge UK, 2004.

Davies, P. C. An overview of the contributions of John Archibald Wheeler. In: Barrow, J. D., Davies, C. W., and Harper, C. L., Jr.—eds. Science and Ultimate Reality: Quantum Theory, Cosmology, and Complexity, pp. 3-19. Cambridge University Press, Cambridge UK, 2004.

Dieter H. (ed): Fundamentals of quantum information. Quantum computation, Communication, Decohernce and all that. Springer, NY. 2002.

Dieter Heiss (ed): Fundamentals of quantum information. Quantum computation, Communication, Decohernce and all that. Springer, NY. 2002.

Dinner, S. The wave-particle duality as an interplay between order and chaos. In: Dinner S., Fargue D., Lochak G., Sellery F—eds. The Wave-Particle Dualism. Riedel Publishing Co., Dordrecht, Holland, 215-229, 1984.

Einstein A., Podolsky B., Rosen N. Can quantum-mechanical description of physical reality be considered complete? Phys. Rev., v. 47, pp. 777-780, 1935.

Einstein, A. From Einstein's letter to Born, quoted by Clark, R. W. Einstein The Life and Time, p. 609. The World Publishing Company, NY, 1971.

Einstein A. The Meaning of Reality. MJF Books, New York, 1984 (first ed. 1922).

Dr. Alexander Yabrov

Einstein, A. Quotable Einstein (A. Caleprice—ed.). Princeton University Press, Princeton, NJ, 1996.

Einstein, A. Quoted from Clark, R. Einstein, the Life and Time. The World Publishing Company, New York, p. 612 (1971).

Einstein, A. to Schrodinger, E.,—Letter of June 1935. Quoted from Moore, W.
"Schrodinger—Life and Thought", p. 304. A. Caleprice. Einstein. Cambridge University Press, Cambridge, MA, 1990.

Einstein, A. and Infeld, L. The Evolution of Physics. Tuchstone, NY, 1937.

Everett, H. "Relative state" formulation of quantum mechanics. pp. 454-462, v.29, No. 3, Review of Modern Physics, 1957.

Feynman R. The Character of Physical Law (MIT Press, Cambridge, MA, 1990).

Fitch, Val L. The discovery of charge conjugation-parity asymmetry. Science, pp. 989-993, v. 212, No. 4498, 1981.

Folk GE. Introduction of Environmental Physiology. Lee and Febierg, Philadelphia, 1966.

Folk GE. Introduction to Environmental Physiology. Lee and Febierg I Philadelphia, 1974.

Folk G.E. Introduction to Environmental Physiology. Lee and Febierg I Philadelphia, 1979.

Ford, K. W. The Quantum World. Quantum Physics for Everyone. Harvard University Press. Cambridge, MA, 2004.

Ford, K. Written review. Princeton University, Princeton, NJ, personal communication, 3.11.05.

Ford K.W. In Love with Flying. H Bar Press. Philadelphia, Pennsylvania, 2007.

Gamov, G. Thirty Years that Shook Physics. Dubleday, Garden City, NY, 1966.

Gell-Mann, M. The Quark and the Jaguar. W. H. Freeman and Company. New York, 1994.

Gould S. Ever Since Darwin: Reflection in Natural History. New York: Penguin; 1977.
Greene, B. The Elegant Universe. Vintage Books. New York, 1999.

Gross DJ. Unified Theories of Everything. Instituto Italiano. Napoly, 1989.

Gullemin, V. The Story of Quantum Mechanics. Charles Scribner's Sons, NY, 1968.

Halpern, P. The Great Beyond. John Wiley & Sons, Inc., Hoboken, NJ, 2004.

Hawking, S. A Unified Theory of Everything. Lecture, Convocation Hall, University of Cambridge University, Cambridge, 27 April, 1980.

Hawking, S. and Penrose, R. The Nature of Space and Time. Princeton Univ. Press, Princeton, 1996.

Heidegger, M. Being and Time. Harper & Row, NY, 1962.
Heisenberg, W. A quotation repeated both by Bohr and Heisenberg: "Clarity is gained through breadth". Physics and Beyond, p. 246, Harper & Raw, Publishers, New York, 1971.

Dr. Alexander Yabrov

Heisenberg, W. Physics and Beyond. Encounters and Conversations. Harper & Row, New York, 1971.

Heisenberg, W. Physics and Philosophy. Hirzel, Leipzig, 1944.

Heisenberg, W. The Physical Conception of Nature. Greenwood Press, Publishers, Westport, Connecticut, 1958.

Hopfield, J. J. Physics, computation and biology. In: Springers Proceedings in Physics. Evolutionary Trends in Physics (eds. M. Suzuki, R. Kubo), pp. 217-227, v. 57, Springer-Verlag, Berlin, 1991.

Isaacs, A., Lindenmann, K. Virus interference. 1. Interferon. Proc. Roy. Soc., 258-267, 147. London, 1957.

Kant, I. Critique of Judgment. Hackett Publishers, New York, 1987 (1790).

Kant, I. Critique of Pure Reason. Humanities Press, New York, 1997.

Keller, E. Making sense of Life: Explaining Biological development with Models, Metaphors, and Machines. Harward University Press, Cambridge, MA, 2003.

Kuhn, T. S. The Structure of Scientific Revolutions. Chicago University Press, Chicago, 2nd ed. 1970 (First ed. 1962).

Laughlin, R. B. and Pines, D. The theory of everything. Proceedings National Academy of Sciences, USA, Jun. 4, 97(1), pp. 28-31, 2000.

Lewontin R. C. Science and Simplicity. The New York Review of Books, pp. 39-42, May 1, 2003.

Lindley, D. The End of Physics. The Myth of a Unified Theory. Basic Books. New York, 1993.

Lochak, G. De Broglie initial conception of de Broglie Waves. In: Dinner, S., Fargue, D., Lochak, G., Sellery, F.—eds.The Wave-Particle Dualism. D. Riedel Publishing Co., Dordecht, Holland, pp. 1-26, 1984.

Lovelock, J. Gaia: A New Look at Life. Oxford University Press, Oxford, 1979.

Lovelock, J. The Ages of Gaia: Biography of Our Living Earth. Norton, NY, 1988.

Lovelock, J., Margulis, L. Atmospheric homeostasis by and for the biosphere: the Gaia hypothesis. Tellus, 26, (1-2), pp. 2-9, 1974.

Mayr E. Darwin's influence on modern thought. Scientific American, 79-83, 2000.

Moore, W. Schrodenger—Life and Thought. Cambridge University Press. Cambridge, 1989.

Newton, I. Principia. Maclehose, Glasgow, 1871.

Okunev Yu.Phase and Phase- Difference Modulation in Digital Communications. Artech House. Boston, MA, 1997.

Oparin. The Origin of Life. New York. Dover, 1952.

Pais, A. (1982) The Science and the Life of Albert Einstein. *Oxford University Press*, NY.

Park, D. "The How and the Why". Princeton University Press, Princeton, NJ (1988).

Dr. Alexander Yabrov

Parker, B. Einstein's Dream. The Search for a Unified Theory of the Universe.
Plenum Press. New York1988.

Peacocke, A. The case for reductionism in the sciences. In: Reductionism in Academic Disciplines (A. Peacocke—ed.). Srhe and Nfer-Nelson, Worester, Gr. Br., 1985.

Penrose, R. Gravity and state vector reduction. In: Quantum Concepts in Space and Time (R. Penrose, C. J. Isham—eds.). Calderon Press, Oxford, p. 129, 1986.

Penrose, R. The Rediscovery of Gravity. In: Farmelo, G. (ed.), It Must be Beautiful. Great Equations of Modern Science. Granta Books. New York, 2002.

Penrose, R. The Road to Reality. Alfred A. Knopf. New York, 2004.

Plato. Complete Works. Hacket Publishing Co., New York, 1997.

Popper K. Realism and the Aim of Science. Rowman and Littlefield, Totowa, NJ, 1983.

Popper, K. Popper Selections (ed. D. Miller). Princeton University Press, Princeton, NJ, 1985.

Popper, K. The Logic of Scientific Discovery. Basic Books, Inc., New York (1959).

Prigogine, From Being to Becoming W. H. Freeman, San Francisco, 1997.

Prigogine, I. The End of Certainty. Time, Chaos, and the new Laws of Nature. The Free Press, New York, 1997.

Requarth, S., Crist, A. Neural cells growth. Scientific American, January, v. 304, No.1, 2011.

Russell, B. The Philosophy of Logical Atomism, pp. 35-155. In: Pears, D., ed. The Philosophy of Logical Atomism. Open Court, La Salle, IL, 1985. (B. Russel—first published 1918).

Schlipp, P.A. Albert Einstein: Philosopher-Scientist. Harper & Brothers Publishers, NY, 1959.

Schrödinger, E. The present situation in quantum mechanics. Naturwissenschaften, 23, 807-812, 1935.

Scholander PF. Studies on man exposed to cold. Fed.Proc. 1054, 17, 1958.

Selye, H. Stress without Distress, McCleland and Srewart, LTD, Toronto, 1974.

Spinoza, B. Dictionary. (D. D. Runes—ed.). Philosophical Library, NY, 1951.

Tarozzi, G. and van de Merwe, A. (eds.). Open Questions in Quantum Physics. D.Riedel, Boston, 1984.

Von Baeyer, H. C. In the beginning was the bit. New Scientist, pp. 26-30, no. 2278, 17 February 2001.

Weinberg S. The revolution that didn't happen. The New York Review of Books, pp. 47-52, October 8, 1998.

Weinberg S. Dreams of a Final Theory. Pantheon Books, NY, 1992.

Weinberg, S. Dream of a Final Theory: The Scientist's Search for the Ultimate Laws of Nature. Vintage, New York, 1994.

Dr. Alexander Yabrov

Weinberg, S. Facing Up. Harvard University Press, Cambridge, MA (2002).

Weinberg, S. On scientific revolutions. New York Review of Books, pp. 11-14, v. XLV, No 15, 1998.

Weinberg, S. The First Three Minutes: A Modern View of the Origin of the Universe. Basic Books, New York, 1977.

Weyl H. Philosiphy of Mathematics and Natural Science. Princeton University Press, Princeton, NJ, 1949.

Wheeler J. A. Geons. v. 97, p. 511, Jan, Phys. Rev., 1955.

Wheeler, J. A. (1957) Assessment of Everett's "relative state" formulation of quantum theory. *Review of Modern Physics*, 29 (3), 463-465.

Wheeler, J. A. (2005) Personal communication. Princeton University, Princeton, NJ. Feb.17.

Wheeler, J. A. At Home in the Universe. AIP Press, Springer, NY, 1996

Wheeler, J. A. Curved empty space as the building material of the physical world: an assessment. In: Logic, Methodology and Philosophy of Science. Proceedings of the 1960 International Congress (E. Nagel, P. Suppes, A. Tarski—eds.). Stanford University Press, 1962.

Wheeler, J. A. Interview to the student newspaper "Engineer". Princeton University, Princeton, NJ, February, 1957.

Wheeler, J. A. Wrestling with the issue: how come existence? In: Frontier Physics (MacDowell, S., Nussenzveig, H.
M., Salmeron, R. A.—editors). World Scientific, River Edge, NJ, 1991.

Wheeler J. A., Ford, K. Geons, Black Holes and Quantum Foam. A Life in Physics. W. W. Norton and Co., NY. 1998.

Whereat AF. Lipid biosynthesis in aortic intima from normal and cholesterol-fed rabbits. Journal of Atherosclerosis Research 4: 272-282, 1964.

Wilczek, F. Four big questions with pretty good answers. In: Buschorn, G. W., Weiss, J. (eds.), Fundamental Physics—Heisenberg and Beyond, pp. 79-98, 2004.

Wilczek F. Fantastic Realities. World Scientific, Singapore, 2006.

Wilczek F. The Lightness of Being. Basic Books, NY, 2008.

Wittgenstein L. Tractatus Logico-Philosophicus. Routlege and Kegan Paul, London, 1961 (First ed. 1921).

Woit P. Not Even Wrong. Basic Books, New York, 2006.

Yabrov, A., On the Mechanism of Cellular Stress. Cytology (Leningrad), II, 2, 137-146, 1969.

Yabrov A. From Uncertainty of Ignorance to Uncertainty of Science. Tractatus Scientifico-Philosophicus. 1st Books Library, Bloomington, Indiana, 2002.

Yabrov A. How Man Exists. 1st Books Library, Bloomington, Indiana, 2001.

Yabrov A. Interferon: Anticancer Agent Having Multifarious Activity. Ariel I (ed.), Yearbook: "Progress in Clinical Cancer", V8, 99-145. Grun and Stratton, New York, 1982.

Dr. Alexander Yabrov

Yabrov A. General Mechanism of Chronic Diseases. Medical Hypotheses 22, 51-87, 1987.

Yabrov A. Theory of Adequate Function of the Organism—the Theoretical Basis for contemporary Medicine. Medical Hypotheses 22, 251-276, 1987a.

Yabrov A. A theory of cellular-function pathology: further development of Virchow's theory of cellular pathology. Medical Hypotheses, 56(4), 434-441, 2001.

Yabrov A. Adequate Function of the Cell: Interaction Between the Needs of the Cell and the Needs of the Organism. Medical Hypotheses, 6, 337-374, 1980a.

Yabrov A. General principle of existence of natural objects. Medical Hypotheses, 19, 3, 331- 337 (1986).

Yabrov A. Interferon and Nonspecific Resistance. Human Sciences Press, New York, 1980.

Yabrov A. Maintenance of adequate function is a general principle of survival of organisms. Medical Hypotheses, 549-574, 5, 1979.

Yabrov A. Theory of adequate function of the organism—the theoretical basis for contemporary medicine. Medical Hypotheses, 251-276, 22, 1987.

Yabrov A. Method of enhancing efficacy of electrical apparatuses. US Patent Office, No. 12/286,257. 2009.

Yabrov A. The Boldest Ideas of Modern Biology. To be published, 2012.

Yabrov A. Relativity of Uncertainty. To be published, 2012a.

Yabrov A., Okunev, Yu. Medicine without drugs—new direction for nano-technology. Medical Hypotheses, 63(1), 149-154, 2004.

Zeilinger, A. On the interpretation and philosophical foundation of quantum mechanics. In: " Vastakohten todellisuus", Ferstschrift for K. V. Laurikainen (U. Ketvel—editor), pp. 1-20, Helsinki University Press, 1996.

Zeilinger, A. Three challenges from John Archibald Wheeler. In: Science and Ultimate Reality: Quantum Theory, Cosmology, and Complexity (Barrow, J. D., Davies, C. W., and Harper, C. L., Jr.—editors), pp. 201-220. Cambridge University Press, Cambridge UK, 2004.

Zeilinger, A. Why the quantum? It from Bit? A participatory universe? Three far-reaching challenges from John Archibald Wheeler and their relation to experiment. In: Science and Ultimate Reality: Quantum Theory, Cosmology, and Complexity (Barrow, J. D., Davies, C. W., and Harper, C. L., Jr.—editors), pp. 201-220. Cambridge University Press, Cambridge UK, 2004.

ABOUT the AUTHOR

Professor Alexander Yabrov—physician (internal and infectious diseases) and biologist-experimenter (medical virology and cell biology). First in the USA, organized industrial production of interferon. All the initial studies of the anticancer effect of interferon approved by the FDA where performed using our preparation. Author of several monographs. Of these—"Interferon and Nonspecific Resistance" (Human Sciences Press, 1980)—The Best Book of the Year in Health Area. Awarded by the American Publishers Association.

Positions in the USSR: Senior Scientist Nuclear Physics Institute, USSR Academy of Sciences; Senior Scientist Institute of Experimental Medicine, USSR Academy of Medical Sciences.

In the USA: Professor Rutgers University; Director R&D of the National Patent Development Corp. and Interferon Sciences, Inc.; President and Director R&D Princeton Biotechnologies, Inc. (present).

www.ingramcontent.com/pod-product-compliance
Lightning Source LLC
Chambersburg PA
CBHW051458170526
45166CB00001B/301